Abnormal Psychology

Abnormal Psychology

Dr. Prem Sunder
PRINCIPAL,
DARSH COLLEGE OF EDUCATION
GOHANA (SONEPAT),
HARYANA

K.S.K. Publishers and Distributors
7/26, Basement, Ansari Road, Darya Ganj,
New Delhi-110002

Published by:

***K.S.K.* Publishers and Distributors**

7/26, Basement, Ansari Road, Darya Ganj, New Delhi-110002.

Ph.: (O) 011-65190343, 43551324, 42564726
(M) 9811088729, (Fax) 011-42564726
(Res.) 47091605

E-mail: *khelsahitya1@rediffmail.com*

Web: *www.khelsahitya.com*

ISBN - 978-81-89983-26-0

PRINTED IN INDIA 2016

Laser Typeset by:
JAIN MEDIA GRAPHICS
Delhi-110035.

Printed by:
ASIAN OFFSET PRINTERS
Delhi

Price: Rs. 700/-

Preface

The phenomena of mental abnormality are both varied and complex. Interest in this field is shared by numerous groups of workers. Proficiency in it requires expertness in several branches of learning not yet adequately organized into any program or curriculum, and skill acquired only through experience.

Educators, hygienists, and all those charged with the care of the young, are deeply concerned with mental abnormalities, from the point of view of avoiding or preventing them. Clergymen and priests, and all who minister to the spiritual life and ethical behavior of mankind have from historic times had a similar interest. The "types" and "characters" of the novel, drama, and short story have also been in close topical contact with this field. Interests such as these may, in general, be called "practical," or "everyday" interests.

Another group of interests is more technical or technological. To physiologist, neurologist, and physician, mental abnormality presents both a professional challenge and a scientific problem. Mental and bodily health are intimately related, to say the least. In many instances, the medical man can, through the medium of the body, "minister to a mind diseased." Medical specialists now adopt some such term as "psychiatry" or "neuro-psychiatry" to designate this province. A slovenly use of the former term extends it indiscriminately to any concern with problems of individual adjustment, such as those of the personnel manager, the parent, social worker, teacher, prison warden, and detective.

This volume is an introduction for students studying abnormal psychology or psychopathology as part of an undergraduate programme in psychology, nursing, or social and behavioural sciences. Under-graduate students who

1

Introduction

Abnormal psychology is an academic and applied subfield of psychology involving the scientific study of abnormal experience and behavior (as in neuroses, psychoses and mental retardation) or with certain incompletely understood states (as dreams and hypnosis) in order to understand and change abnormal patterns of functioning. The definition of what constitutes 'abnormal' has varied across time and across cultures, and varies among individuals within cultures. Today, persistent abnormal functioning is often associated with a mental disorder. In general, abnormal psychology studies people who are consistently unable to adapt and function effectively in a variety of conditions. An individual's ability to adapt and function can be affected by a number of variables, including one's genetic makeup, physical condition, learning and reasoning, and socialization.

People have tried to explain and control abnormal behavior for thousands of years. Historically, there have been three main approaches to abnormal behavior: the supernatural, biological, and psychological traditions. In the supernatural tradition, abnormal behaviors are attributed to agents outside human bodies. According to this model, abnormal behaviors are caused by demons, spirits, or the influences of moon, planets, and stars. During the Stone Age, trepanning was performed on those who had mental illness to literally cut the evil spirits out of the victim's head. The Ancient Chinese, Egyptians, and Hebrews, on the other hand, believed that these evil demons were punishments from God, and advocated exorcism. By the time of the Greeks and Romans, mental illnesses were thought to be caused by an imbalance of the four humors, leading to draining of fluids from the brain. During the Dark Ages, many Europeans believed that the power of witches, demons, and spirits caused abnormal behaviors. People with psychological disorders were thought to be possessed by evil spirits that had to be exorcised through religious rituals. If exorcism failed, some authorities advocated steps such as confinement, beating, and other types of torture to make the body uninhabitable by witches, demons, and spirits. The belief that witches, demons, and spirits are responsible for the abnormal behavior continued into the 15th century. Swiss alchemist, astrologer, and physician Paracelsus (1493-1541), on the other hand, rejected the idea that abnormal behaviors were caused by witches, demons, and spirits and suggested that people's mind and behaviors were influenced by the movements of the moon and stars.

This tradition is still alive today. Some people, especially in the developing countries and some followers of religious sects in the developed countries, continue to believe that supernatural powers influence human behaviors. However, this tradition has been largely replaced in Western academia by the biological and psychological traditions. In the biological tradition, psychological disorders are attributed to biological causes and in the psychological tradition, disorders are attributed to faulty psychological development and to social context.

The Greek physician Hippocrates, who is considered to be the father of Western medicine, played a major role in the biological tradition. Hippocrates and his associates wrote the Hippocratic Corpus between 450 and 350 BC, in which they suggested that abnormal behaviors can be treated like any other disease. Hippocrates viewed the brain as the seat of consciousness, emotion, intelligence, and wisdom and believed that disorders involving these functions would logically be located in the brain. These ideas of Hippocrates and his associates were later adopted by Galen, the Roman physician. Galen extended these ideas and developed a powerful and influential school of thought with in the biological tradition that extended well into the 18th century.

CORE CONCEPTS: EXPLAINING ABNORMAL BEHAVIOUR

Abnormal psychology consists of three core concepts: cultural and historical relativism, the principle of multiple causality and the connection between mind and body.

Cultural and Historical Relativism

Throughout time, societies have proposed several explanations of abnormal behavior within human beings. Beginning in some hunter-gatherer societies, animists have believed that people demonstrating abnormal behavior are possessed by malevolent spirits. This idea has been associated with trephination, the practice of cutting a hole into the individual's skull in order to release the malevolent spirits.

A more formalized response to spiritual beliefs about abnormality is the practice of exorcism. Performed by religious authorities, exorcism is thought of as another way to release evil spirits who cause pathological behavior within the person. In some instances, individuals exhibiting unusual thoughts or behaviors have been exiled from society or worse. Perceived witchcraft, for example, has been punished by death. Two Catholic Inquisitors wrote a manual, the Malleus Maleficarum, that became co-opted by many Inquisitors and witch-hunters. It contained an early taxonomy of deviant behavior and proposed guidelines for prosecuting deviant individuals.

The act of placing mentally ill individuals in a separate facility known as an asylum dates to 1547, when King Henry

VIII of England established the St. Mary of Bethlehem asylum. Asylums remained popular throughout the Middle Ages and the Renaissance era.

Multiple Causality

The number of different theoretical perspectives in the field of psychological abnormality has made it difficult to properly explain psychopathology. The attempt to explain all mental disorders with the same theory leads to reductionism (explaining a disorder or other complex phenomena using only a single idea or perspective). Most mental disorders are composed of several factors, which is why one must take into account several theoretical perspectives when attempting to diagnose or explain a particular behavioral abnormality or mental disorder. Explaining mental disorders with a combination of theoretical perspectives is known as multiple causality.

The diathesis-stress model emphasizes the importance of applying multiple causality to psychopathology by stressing that disorders are caused by both precipitating causes and predisposing causes. A precipitating cause is an immediate trigger that instigates a person's action or behavior. A predisposing cause is an underlying factor that interacts with the immediate factors to result in a disorder. Both causes play a key role in the development of a psychological disorder.

Mind and Body

A paradigm is a general viewpoint on the world and is much broader than a theory. Today's field of psychology revolves around two major paradigms for explaining mental disorders, the *psychological* paradigm and the *biological* paradigm. The psychological paradigm focuses more on the humanistic, cognitive and behavioral perspectives. The biological paradigm includes the theories that rely more on physical causes such as genetics and neurochemistry.

Recent Concepts of Abnormality

- *Statistical abnormality* - when a certain behaviour/characteristic is relevant to a low percentage of the population. However, this does not necessarily mean that such individuals are suffering from mental illness (for example, statistical abnormalities such as extreme wealth/attractiveness)
- *Psychometric abnormality* - when a certain behaviour/characteristic differs from the population's normal dispersion e.g. having an IQ of 35 could be classified as abnormal, as the population average is 100. However, this does not specify a particular mental illness.
- *Deviant behaviour* - this is not always a sign of mental illness, as mental illness can occur without deviant behaviour, and such behaviour may occur in the absence of mental illness.
- *Combinations* - including distress, dysfunction, distorted psychological processes, inappropriate responses in given situations and causing/risking harm to oneself.

Approaches

- *Somatogenic* - abnormality is seen as a result of biological disorders in the brain. However, this approach has led to the development of radical biological treatments e.g. lobotomy.
- *Psychogenic* - abnormality is caused by psychological problems. Psychoanalytic (Freud), cathartic, hypnotic and humanistic psychology(Carl Rogers, Abraham Maslow) treatments were all derived form this paradigm. This approach has, as well, led to some esoteric treatments: Mesmer used to place his patients in a darkened room with music playing, then enter it wearing a flamboyant outfit and poke the 'infected' body areas with a stick.

CLASSIFICATION

DSM-IV TR

The standard abnormal psychology and psychiatry reference book in North America is the Diagnostic and Statistical Manual of the American Psychiatric Association. The current version of the book is known as DSM IV-TR. It lists a set of disorders and provides detailed descriptions on what constitutes a disorder such as Major Depressive Disorder or anxiety disorder. It also gives general descriptions of how frequent the disorder occurs in the general population, whether it is more common in males or females and other such facts. The diagnostic process uses five dimensions called 'axes' to ascertain symptoms and overall functioning of the individual. These axes are as follows——

- *Axis I - Particular clinical syndromes*
- *Axis II - Pervasive disorders (Personality Disorders, Mental Retardation)*
- *Axis III - General medical conditions*
- *Axis IV - Psychosocial/environmental problems*
- *Axis V - Global assessment of functioning (often referred to as GAF)*

ICD-10

The major international nosologic system for the classification of mental disorders can be found in the most recent version of the International Classification of Diseases, 10th revision (ICD-10). The ICD-10 has been used by World Health Organization (WHO) Member States since 1994. Chapter five covers some 300 "Mental and behavioural disorders." The ICD-10's chapter five has been influenced by APA's DSM-IV and there is a great deal of concordance between the two. WHO maintains free access to the ICD-10 Online . Below are the main categories of disorders:

- *F00-F09* Organic, including symptomatic, mental disorders
- *F10-F19* Mental and behavioural disorders due to psychoactive substance use
- *F20-F29* Schizophrenia, schizotypal and delusional disorders
- *F30-F39* Mood [affective] disorders
- *F40-F48* Neurotic, stress-related and somatoform disorders
- *F50-F59* Behavioural syndromes associated with physiological disturbances and physical factors
- *F60-F69* Disorders of adult personality and behaviour
- *F70-F79* Mental retardation
- *F80-F89* Disorders of psychological development
- *F90-F98* Behavioural and emotional disorders with onset usually occurring in childhood and adolescence
- *F99* Unspecified mental disorder

Etiology

Genetics

- Investigated through family studies, mainly of monozygotic (identical) and dizygotic (fraternal) twins, often in the context of adoption.
- These studies allow calculation of a *heritability coefficient*.

Biological Causal Factors

- Neurotransmitter [Imbalances of Neurotransmitters like (1) Norepinephrine (2) Dopamine (3) Serotonin and (4) GABA (Gamma aminobutryic acid)]and Hormonal imbalances in the brain
- Genetic vulnerabilities
- Constitutional liabilities [Physical Handicaps and temperament]
- Brain dysfunction and neural plasticity

- Physical deprivation or disruption [Deprivation of basic physiological needs]

PSYCHOLOGICAL FACTORS

- Psychoanalysis (Freud)
- Behavioural therapy (Wolpe) based on behaviourism, and involving classical and operant conditioning.
- Humanistic therapy aiming to achieve self-actualisation (Carl Rogers, 1961)
- Cognitive Behavioural Therapy aims to influence thought and cognition (Beck, 1977).

SOCIO-CULTURAL FACTORS

- Effects of urban/rural dwelling, gender and minority status on state of mind.
- British Psychiatric Morbidity Survey conducted by Jenkins (1998)

SYSTEMIC FACTORS

- Family systems
- Negatively Expressed Emotion playing a part in schizophrenic relapse and anorexia nervosa.

BIOPSYCHOSOCIAL FACTORS

- Holistic causal model
- Illness dependent on stress 'triggers'.

So we can conclude that the definition of the word abnormal is simple enough: deviating from the norm. However, applying this to psychology poses a complex problem: what is normal? Whose norm? For what age? For what culture? Some would simply classify what is "good" as normal and what is "bad" as abnormal, but this is a vague and narrow definition and brings up many of the same questions for the definition of "good" as does the definition for "normal". There are many more ways of determining a more objective reference point. One way of defining abnormality is statistical deviation. Most human characteristics are nicely distributed along a smooth bell-shaped curve.

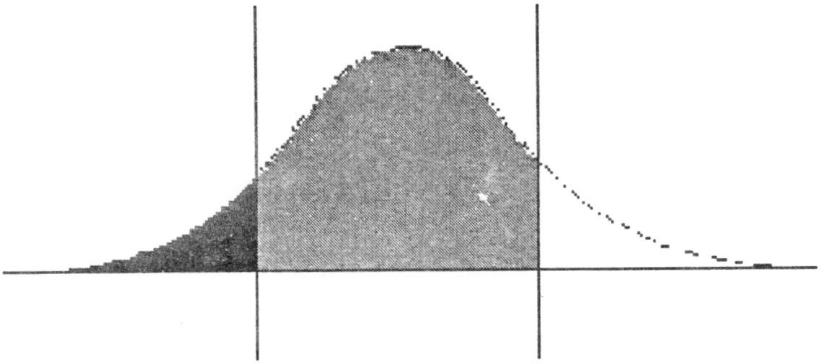

Those who stray too far from the average on this curve are then considered abnormal. This definition, however, has its problems. It does not recognize valuable derivations, such as genius, nor does it recognize common but maladaptive behavior like smoking and drinking. Nonetheless, this method is both objective and scientific. A very simple idea that can be used to classify abnormal behavior is personal distress. Basically, if a person is content with their life, then they are of no concern to the mental health field. However, if a person's thoughts or behaviors are causing them personal discomfort or unhappiness, then they will be considered abnormal.

The most common criterion for defining abnormality, however, is maladaptiveness. There are two aspects of maladaptive bahvior:

1. *Maladaptive to one's self* - inability to reach goals, to adapt to the demands of life and
2. *Maladaptive to society* - interferes, disrupts social group functioning.

This type of definition allows much flexibility. It provides room for conforming behavior to society's norms as well as deviant behavior as long as it is not self-damaging. It makes abnormal the relative term it needs to be, dependent upon each individual's life and circumstances. There are certain catergories of behavior that suggest the presence of psychological disorders which are, in one way or another, maladaptive in that they threaten the well-being of the individual. These catergories include long periods of discomfort, impaired functioning, bizarre behavior, and disruptive behavior.

Long Periods of Discomfort

Given, everyone experiences some kind of psychological discomfort during their life. This could be anything as simple as worrying about a calculus test to grieving the death of a loved one. This distress, however, is related to a real, related, or threatened events and passes away with time. When such distressing feelings, however, persist for an extended period of time and seem to be unrelated to events surrounding the person, they would be considered abnormal and could suggest a psychological disorder.

Impaired Functioning

Here, again, there must be made a distinction between simply a passing period of inefficiency and prolonged inefficiency which seems unexplainable. For instance, a very brilliant person who consistently fails classes or someone constantly changing jobs for no reason.

Bizarre Behavior

There are many things people do that others would find strange. The various piercings today's younger generation chooses to get and their style of dress may seem bizarre to adults, but their motivations are not hard to understand, which keeps them from being considered clinically abnormal. Bizarre behavior that has no rational basis, however, seems to indicate that the individual is confused. The psychoses frequently bring on hallucinations (baseless sensory perceptions) or delusions (beliefs which are patently false yet held as truth by the individual).

Disruptive Behavior

Disruptive behavior means impulsive, apparently uncontrollable behavior that disrupts the lives of others or deprives them of their human rights on a regular basis. This type of behavior is characteristic of a severe psychological disorder. An example of this is the antisocial personality disorder. All of these types of behavior are maladaptive because they directly affect the well-being of the individual and those around them, and block the growth and fulfillment of the individual's potential.

2

Childhood Behaviour Disorders

A wide variety of psychological problems may occur in childhood. These include problems that compromise children's capacities to learn and communicate, such as intellectual disability; language delay; specific learning disabilities; and pervasive developmental disorders including autism. Problems developing bowel and bladder control, sleeping and waking routines, and feeding and eating disorders such as anorexia nervosa may also occur in childhood and adolescence. Children and adolescents may develop neuropsychological problems and adjustment difficulties secondary to conditions such as epilepsy or head injury. All of these difficulties are of concern to psychologists who study abnormal behaviour (Carr, 1999). However, in addition to these difficulties, two broad classes of conditions have been a focus for psychologists who study abnormal behaviour in childhood. These are disruptive behaviour disorders (such as attention deficit hyper-activity disorder, oppositional defiant disorder, conduct disorder) and emotional disorders (such as anxiety and depression).

Later the emotional disorders—anxiety and depression—in children, adolescents and adults will be addressed. In this chapter, disruptive behaviour disorders will be the central focus for three main reasons. First, disruptive behaviour disorders in children and adolescents are particularly prevalent in the community. Second, these disorders are among the most common referrals to child mental health services. Third, in the long term if left untreated, these disorders are extremely costly both to the children who suffer from them and to society. The principal clinical features of attention deficit hyperactivity disorder, oppositional defiant disorder and conduct disorder are given in Table below. It is noteworthy that all three of these conditions entail behaviour that is troublesome for others as well as for the child.

After considering the clinical features and epidemiology of attention deficit hyperactivity disorder, oppositional defiant disorder and conduct disorder, theoretical explanations for these problems are presented in this chapter. Each of these specific explanations has been developed within the context of one of four broad theories. These are the biological, psychodynamic, cognitive-behavioural and family systems theories of psychological problems. Later these four broad theories are reviewed with reference to their main attributes, their contributions to our understanding and treatment of psychological problems, and their limitations.

Attention Deficit Hyperactivity Disorder

Attention deficit hyperactivity disorder, attention deficit disorder, hyperkinetic disorder, hyperkinesis and minimal brain dysfunction are some of the terms used for a syndrome characterized by persistent overactivity, impulsivity and difficulties in sustaining attention (Hinshaw, 1994; Taylor, 1994a; Barkley, 1998). In this chapter preference will be given to the term attention deficit hyperactivity disorder (ADHD) since this is currently the most widely used.

Case Example

Timmy, aged 6, was referred for assessment because his teachers found him unmanageable. He was unable to sit still in school and concentrate on his schoolwork. He left his chair frequently and ran around the classroom shouting. This was distracting for both his teachers and classmates. Even with individual tuition he could not apply himself to his schoolwork. He also had difficulties getting along with other children. They disliked him because he disrupted their games. He rarely waited for his turn and did not obey the rules. At home he was consistently disobedient and according to his father ran 'like a motorboat' from the time he got up until bedtime. He often climbed on furniture and routinely shouted rather than talked.

Family history. Timmy came from a well-functioning family. The parents had a very stable and satisfying marriage and together ran a successful business.

Table.

Clinical features of childhood behaviour disorders

Domain	*ADHD*	*Oppositional defiant disorder*	*Conduct disorder*
Cognition	• Short attention span • Distractibility • Unable to foresee consequences of behaviour • Immature self-speech (internal language) • Low self-esteem • Lack of conscience	• Limited internalization of social rules or norms • Hostile attributional bias	• Limited internalization of social rules or norms • Hostile attributional bias

(Cont...)

	• Learning difficulties and poor school performance • Lack of impulse control		
Affect	• Excitability • Low frustration tolerance • Low mood	• Anger and irritability	• Anger and irritability
Behaviour	• High rate of activity • Delay in motor development and poor co-ordination • High level of risk-taking behaviour	• Persistent pattern of defiance towards adults in authority • Aggression • Temper tantrums	• Persistent broad pattern of antisocial behaviour • Defiance • Aggression • Destructiveness • Deceitfulness and theft • Cruelty • Truancy • Running away • Coercive sex • Preadolescent drug use
	• Immature physical size and bone growth		• Physical problems

(Cont.)

		associated with risk-taking behaviour such as fighting, drug abuse or casual unsafe sex
Physical condition	• Minor physical abnormalities • Allergies • Increased respiratory	

Their daughter, Amanda, was a well-adjusted and academically able 8-year-old. The parents were careful not to favour the daughter over her brother or to unduly punish Timmy for his constant disruption of his sister's activities. However, there was a growing tension between each of the parents and Timmy. While they were undoubtedly committed to him, they were also continually suppressing their growing irritation with his frenetic activity, disobedience, shouting and school problems. Within the wider family there were few resources that the parents could draw on to help them cope with Timmy. The grandparents, aunts and uncles lived in another county and so could not provide regular support for the parents. Furthermore, they were bewildered by Timmy's condition, found it very unpleasant and had gradually reduced their contact with Timmy's nuclear family since his birth.

Psychometric Assessment and Child Interview. Psychometric evaluation showed that his overall IQ was within the normal range but Timmy was highly distractible and had literacy and numeracy skills that were significantly below his overall ability level. Timmy perceived himself to be a failure. He believed that he could not do anything right at home or at school and he was sad that the other children did not want to play with him. He believed that his teacher disliked him and doubted his parents' love for him.

Developmental History. There were a number of noteworthy features in Timmy's developmental history. He had suffered anoxia at birth and febrile convulsions in infancy. He had also had episodes of projectile vomiting. His high activity level and demandingness were present from birth. He also displayed a difficult temperament, showing little regularity in feeding or sleeping; intense negative emotions to new stimuli; and was slow to soothe following an intense experience of negative emotion.

Formulation. Timmy was a 6-year-old boy with home- and school-based problems of hyperactivity, impulsivity and distractibility of sufficient severity to warrant a diagnosis of attention deficit hyper-activity disorder. Possible predisposing factors included anoxia at birth, subtle neurological damage due to febrile convulsions in infancy, and a difficult temperament. In Timmy's case ADHD had led to academic attainment difficulties; peer relationship problems; and tension within the family. This wider constellation of difficulties underpinned Timmy's diminishing self-esteem which in turn exacerbated his problems with attainment, peer relationships and family relationships. The absence of an extended family support system for the parents to help them deal with Timmy's difficulties was also a possible maintaining factor. Important protective factors in this case were the commitment of the parents to resolving the problem and supporting Timmy, and the stability of Timmy's nuclear family.

Treatment. Treatment in this case involved both psycho-social and pharmacological intervention. The psychosocial intervention included parent and teacher education about ADHD; behavioural parent training; self-instructional training for the child; a classroom-based behavioural programme, and provision of periodic relief care/holidays with specially trained foster parents. Timmy was also given stimulant therapy, specifically a twice-daily dose of methylphenidate.

Clinical Features

The clinical features of ADHD in the domains of cognition, affect, behaviour, physical health and interpersonal adjustment are given in Table above. Timmy, in the case example, showed all of these. With respect to cognition, short attention span, distractibility and an inability to foresee the consequences of action are the main features. There is usually a poor internalization of the rules of social conduct and in some instances low self-esteem may be present. With respect to affect, excitability associated with lack of impulse control is the dominant emotional state. This may be coupled with depressed mood associated with low self-esteem in some cases. With ADHD it is the high rate of activity, common co-morbid aggressive antisocial behaviour, excessive risk-taking and poor school performance associated with inattention that is the cardinal behavioural features. With respect to physical health in ADHD, in some instances food allergies may be present.

Injuries or medical complications associated with antisocial behaviour such as fighting and drug-abuse may also occur. Relationship difficulties with parents, teachers and peers are the principal interpersonal adjustment problems. Difficulties with turn-taking in games due to impulsivity make children with ADHD poor playmates. The failure of children with ADHD to internalize rules of social conduct at home and to meet parental expectations for appropriate social and academic behaviour leads to conflictual parent-child relationships. In school, youngsters with ADHD pose classroom management problems for teachers, and these children invariably have problems benefiting from routine teaching and instructional methods. For these reasons, their relationships with teachers tend to be conflictual.

Historically, a narrow definition of ADHD has been included in the ICD classification system which is widely used in the UK, with great emphasis being placed on the stability of the overactivity problems across home and school contexts. In contrast, in the US, this cross-situation stability has not been a core diagnostic criterion within early editions of the DSM (Hinshaw, 1994). In view of this historical difference in diagnostic practices, it is particularly noteworthy that currently in both the North American DSM IV and the ICD 10 which is used widely in Europe, it is stipulated that symptoms must be present in two or more settings such as home and school for a positive diagnosis of ADHD to be made.

Epidemiology

Reviews of epidemiological studies of ADHD report overall prevalence rates varying from 1 to 19 per cent depending upon the stringency of the diagnostic criteria applied and the demographic characteristics of the populations studied (Hinshaw, 1994). Using DSM IV criteria a prevalence rate of about 3-5 per cent has been obtained. The prevalence of ADHD varies with gender and age. ADHD is more prevalent in boys than girls and in preadolescents than in late adolescents. Comorbidity for conduct disorder and ADHD is about 20 per cent in community populations and possibly double this figure in clinical populations. Comorbidity for emotional disorders, such as anxiety or depression, and ADHD is about 10 per cent in community populations. In clinical populations the comorbidity rate is maybe twice this figure. Virtually all children with ADHD have attainment problems. However, comorbid severe specific learning difficulties have been estimated to occur in 10-25 per cent of cases. A proportion of youngsters with ADHD have comorbid developmental language delays and elimination problems although reliable epidemiological data are unavailable.

About a third of children with ADHD have a good prognosis, about a third have a moderate prognosis and a third have a poor prognosis (Hinshaw, 1994). For two-thirds of cases, the primary problems of inattention, impulsivity and hyperactivity persist into late adolescence and for some of these the primary symptoms persist into adulthood.

Roughly a third develop significant antisocial behaviour problems in adolescence including conduct disorder and substance abuse, and for most of this subgroup these problems persist into adulthood leading to criminality. Occupational adjustment problems and suicide attempts occur in a small but significant minority of cases.

Etiological Theories

Biological, cognitive-behavioural and family systems theories have been developed to explain the etiology and maintenance of symptomatology in ADHD.

Biological Theories

Biological theories which focus on the role of genetic factors, structural brain abnormalities, neurotransmitter dysregulation, dietary factors and hypo-arousal have guided much research on the etiology of ADHD.

Genetic hypotheses. Genetic theories suggest that a predisposition to hyperactivity is inherited by children who develop ADHD. Twin and family studies support the view that genetic factors play an important role in determining temperamental activity levels in the normal population. However, other environmental factors (either intrauterine or psychosocial) would be required in addition to a genetic predisposition to high activity levels to account for the development of the clinical syndrome of ADHD (Stevenson, 1992; Hinshaw, 1994). It may be that, in some cases, temperamentally overactive children sustain a prenatal or early childhood neurological insult and go on to develop ADHD, whereas others with an overactive temperament develop the syndrome following participation in particular non-optimal types of parent-child interaction. For a small subgroup of children with ADHD, the syndrome appears to be caused by a genetic condition resulting in a generalized resistance to thyroid hormone (Hauser *et al.*, 1993).

Organic deficit hypothesis. Early work on ADHD was premised on the hypothesis that the syndrome reflected an organic deficit: probably some form of minimal brain damage (Strauss and Lehtinen, 1947). Sophisticated neuroimaging studies have failed to reveal a specific structural brain abnormality which typifies cases of ADHD, and neuropsychological studies have failed to reveal a unique pattern of cognitive deficits associated with either localized or diffuse brain damage characteristic of youngsters with ADHD (Tannock, 1998; Barkley *et al.*, 1992a). However, a number of factors which might lead to brain damage during the prenatal or perinatal periods are more prevalent among youngsters with ADHD than normal controls. These include: prenatal difficulties, maternal smoking during pregnancy, maternal alcohol use during pregnancy, low foetal heart rate during delivery, small head circumference at birth, minor physical abnormalities, low birth weight, a high rate of diseases of infancy, lead poisoning and early neurological insult or brain damage (Taylor, 1994a; Cantwell and Hannah, 1989; Barkley, 1998). It is important to point out that these factors which may contribute to the development of an organic deficit are not unique to ADHD and occur also in youngsters with other disorders. Therefore they probably interact with other factors in contributing to the development of ADHD.

Neurotransmitter dysregulation hypothesis. The neurotransmitter dysregulation hypothesis attributes the symptoms of ADHD to abnormalities in neurotransmitter functioning at the synapses affected by psychostimulants which ameliorate the symptomatology of ADHD. McCracken (1991) has shown that a dysregulation of the dopamine system in the ventral tegmental areas of the brain and norepinephrine and epinephrine systems in the locus coeruleus may be present in ADHD and it is probably these systems that are influenced by effective stimulant therapy with drugs such as methylphenidate and dextroamphetamine. Approximately 60-90 per cent of children with ADHD respond to these stimulants. They show a reduction in symptomatology and an improvement in both academic and social functioning, although positive effects dissipate when treatment ceases (Taylor, 1994b; Hinshaw, 1994; Gadow, 1992).

Dietary hypotheses. The dietary hypothesis attributes the symptoms of ADHD to children's reaction to certain features of their daily diet. Originally Feingold (1975) argued that artificial food additives such as colourants accounted for a substantial proportion of ADHD symptomatology. However, controlled trials of additive-free diets did not support his position (Taylor, 1994a). Egger *et al.*, (1985) refined Feingold's original allergy theory and argued that particular children with ADHD may have unique allergy profiles and if their diet is modified so as to exclude the precise substances to which they are allergic, then their activity and attention problems may improve. Carefully controlled dietary studies have supported Egger's theory (Egger *et al.*, 1985).

Hypo-arousal hypotheses. The hypo-arousal hypothesis explains hyperactivity and inattention as a failure to be sufficiently aroused by signal stimuli to attend to them and regulate activity levels. Hyperactivity may also, according to this hypothesis, reflect stimulus-seeking behaviour. Psychophysiological studies show that ADHD children show reduced psychophysiological responsiveness (as assessed by electroencephalograph (EEG), galvanic skin response (GSR) and heart rate) to novel stimuli with signal value. However, similar unresponsiveness characterizes children with learning disorders and conduct disorders (Taylor, 1994a). The use of vivid stimuli in academic settings and highly salient and immediate reinforcers is implicated by the hypo-arousal hypothesis. Reward systems and operant programmes conforming to these specifications have been found to have significant short-term effects (Pelham and Hinshaw, 1992).

Cognitive-behavioural Theories

A number of theories which highlight the importance of deficits in specific cognitive or behavioural processes as the central factor underlying ADHD symptomatology have been proposed. Four of these will be mentioned below. All attempt to show how the overall syndrome of inattention, overactivity and impulsivity may be accounted for by a single underlying core deficit, be it one of the three core symptoms of ADHD or some other cognitive or behavioural process.

Inattention hypothesis. The attentional deficit hypothesis argues that problems with sustaining attention on a single task and screening out other distracting stimuli are the core difficulty that underpins the other symptoms of impulsivity and overactivity in ADHD (e.g. Douglas, 1983). That is, youngsters with ADHD at the outset of a task requiring attention will perform at a level equivalent to normal children but, over time, they will show more errors which are directly attributable to the inability to sustain attention. This problem with sustaining attention leads them to change the focus of their attention frequently and this is manifested at a behavioural level as excessive impulsivity and overactivity. On certain laboratory tasks children with ADHD show a gradual deterioration in sustained attention. However, on other tasks they show immediate selective attention problems compared to normals and they also display overactivity while asleep (Hinshaw, 1994; Taylor, 1994a). These findings suggest that a deficit in sustained attention alone cannot fully account for the ADHD syndrome.

Hyperactivity hypothesis. The hyperactivity hypothesis argues that a problem with inhibiting motor activity is the core deficit that underpins the ADHD syndrome and can account for inattention and impulsivity (e.g. Schachar, 1991). There is a large body of evidence which shows that hyperactivity is unique as a symptom to children with ADHD compared to children with other psychological problems and that hyperactivity as a construct correlates with many academic indices of attentional problems (Hinshaw, 1994; Taylor, 1994a).

Impulsivity hypothesis. This hypothesis argues that a core problem in inhibiting cognitive and behavioural responses to specific stimuli leads to poor performance on tasks apparently requiring good attentional abilities and also on tasks requiring careful regulation of behaviour. Thus the central problem in ADHD according to this hypothesis is with cognitive and behavioural impulsivity or disinhibition (e.g. Barkley, 1994; Schachar and Logan, 1990). According to this theory, with academic tasks apparently requiring high levels of sustained attention, children with ADHD have problems using systematic cognitive problem-solving strategies because they are cognitively impulsive. Also, in both academic and social situations, children with ADHD engage in careless work practices in school and engage in socially inappropriate behaviour with peers, parents and teachers because they are behaviourally impulsive. There is some evidence to show that, while children with ADHD may know and understand problem-solving skills and social skills, they fail to use them appropriately in academic and social situations (Hinshaw, 1996; Pelham and Hoza, 1996; Abikoff and Hechtman, 1996).

Rule-following deficit hypothesis. Barkley (1981) posits a difficulty in rule-following as the central problem in ADHD. This rule-following deficiency is due to an inability to use inner and outer speech as a discriminative stimulus to cue particular responses. In addition to the extensive evidence for developmental language delay in children with ADHD, there is some support for the delay in the development of internal speech in children with this condition (Berk and Potts, 1991).

To compensate for deficits in attention, regulation of motor activity, impulsivity and rule-following, various skills training programmes have been developed, largely within the cognitive-behavioural tradition. Self-instructional training for managing academic tasks and social skills training to manage relationship problems (particularly those involving peers) are the main types of skills taught within these programmes. Because of the negligible impact that such programmes have when conducted in isolation, they are now offered as one element of a multimodal package involving stimulant medication, family intervention and school intervention (Abikoff and Hechtman, 1996; Pelham and Hoza, 1996; Hinshaw, 1996). Hinshaw (1994) has argued that it is probable that compensatory problem-solving and social skills training programmes can have optimum effects when offered after a consistent home- and school-based contingency management programme has been established and stimulant treatment is in progress.

Family Systems Theory

Family systems theories have focused largely on the role of the family system or the wider social context in the etiology and maintenance of ADHD. With respect to family problems, parental psychological problems such as depression, aggression or alcohol abuse, exposure to marital discord, overintrusive parenting during infancy, and coercive parent-child interactions in childhood and adolescence have all been found to have associations with ADHD (Hinshaw, 1994; Taylor, 1994a; Anastopoulos *et al.*, 1996). With respect to the wider social system the following factors have been found to be associated with ADHD: low socio-economic status, institutional upbringing, peer relationship problems and relation-ship problems with school staff (Taylor, 1994a; Barkley, 1998). A problem with much of the research on psychosocial factors in the etiology and maintenance of ADHD is the fact that in many cases comorbid conduct disorders are present, and the risk factors that are identified, which bear a close resemblance to those identified for conduct disorders, may primarily be associated with the etiology of conduct problems rather than ADHD. A second difficulty is untangling the causal chain and establishing which family and relationship difficulties precede the development of

ADHD and are predisposing factors; and distinguishing these from relationship difficulties that evolve in response to ADHD and possibly maintain or exacerbate the condition.

Family-based interventions and multisystemic intervention programmes involving the child's wider social network have evolved from family systems theories of ADHD. These programmes focus on improving parenting skills and enhancing the child's relationships with members of the family and the wider network. Such programmes have been shown to have positive short-term effects on both symptomatology and social adjustment (Anastopoulos *et al.*, 1996; Barkley *et al.*, 1992b).

Intervention

There is a growing consensus within the field that single-factor theories are unlikely to be able to explain the complex and heterogeneous population of youngsters who qualify for a diagnosis of ADHD (Hinshaw, 1994; Barkley, 1998). It is probable that a variety of biological and psychosocial factors interact in complex ways to give rise to the syndrome and that problems with a number of psychological processes, particularly those involved in regulating both cognitive and motor responses, underpin symptomatology. The symptomatology is probably partially maintained and exacerbated by problematic relationships within the family, the peer group and the school. In view of this integrative formulation, it is not surprising that multimodal treatment packages that include behavioural parent training, self-instructional training and school-based contingency management combined with stimulant therapy have been found to be most effective (Nolan and Carr, 2000). With behavioural parent training, parents are coached in how to consistently reinforce socially appropriate behaviour and extinguish inappropriate behaviour. School-based contingency management programmes extend this type of approach into the child's classroom environment. In self-instructional training, children with ADHD are coached in the use of self-instruction to control the way in which they deploy their attention and control their impulses to engage in high levels of activity. With stimulant therapy, children receive regular doses of methylphenidate or dextroamphetamine.

Conduct Disorder and Oppositional Defiant Disorder

Conduct problems constitute a third to a half of all clinic referrals, and chronic conduct problems are the single most costly disorder of adolescence for three reasons (Kazdin, 1995). First, they are remarkably unresponsive to traditional individual approaches to treatment. Positive outcome rates for routine treatments range from 20 to 40 per cent. Second, about 60 per cent of adolescents with conduct problems have a poor prognosis.

Adolescents with chronic conduct disorder turn to adult criminality and develop antisocial personality disorders, alcohol-related problems and a variety of psychological difficulties. They also have more problems with health, educational attainment, occupational adjustment, marital stability and social integration. The third reason for the high cost of conduct problems is the fact that they are intergenerationally transmitted. Adults with a history of conduct disorder rear children with a particularly high prevalence of conduct difficulties.

Case Example

Bill, aged 11, was referred by his social worker for treatment following an incident in which he had assaulted neighbours by climbing up onto the roof of his house and throwing rocks and stones at them. He also had a number of other problems, according to the school headmaster, including academic underachievement, difficulty in maintaining friendships at school and repeated school absence. He smoked, occasionally drank alcohol, and stole money and goods from neighbours. His problems were longstanding but had intensified in the six months preceding the referral. At that time his father, Paul, was imprisoned for raping a young girl in the small rural village where the family lived.

Family History. Bill was one of five boys who lived with their mother, Rita, at the time of the referral. The family lived in relatively chaotic circumstances. Prior to Paul's imprisonment, the children's defiance and rule-breaking, particularly Bill's, were kept in check by their fear of physical punishment from their father. Since his incarceration, there were few house rules and these were implemented inconsistently, so all of the children showed conduct problems, but Bill's were by far the worst. Rita had developed intense coercive patterns of interaction with Bill and John (the second eldest). In addition to the parenting difficulties, there were also no routines to ensure bills were paid, food was bought, washing was done, homework completed or regular meal and sleeping times were observed. Rita supported the family with welfare payments and money earned illegally from farm-work. Despite the family chaos, she was very attached to her children and would sometimes fake them to work with her rather than send them to school because she liked their company.

At the preliminary interview, Rita said that 'her nerves were in tatters'. She was attending a psychiatrist intermittently for pharma-cological treatment of depression. She had a longstanding history of conduct and mood regulation problems, beginning early in adolescence. In particular she had conflictual relationships with her mother and father which were characterized by coercive cycles of interaction. In school she had academic difficulties and peer relationship problems.

Rita had been ostracized by her own family when she married Paul whom they saw as an unsuitable partner for her, since he had a number of previous convictions for theft and assault. Paul's family never accepted Rita, because they thought she had 'ideas above her station'. Rita's and Paul's parents were in regular conflict, and each family blamed the other for the chaotic situation in which Paul and Rita had found themselves. Rita was also ostracized by the village community in which she lived. The community blamed her for driving her husband to commit rape.

Paul, the father, also had longstanding difficulties. His conduct problems began in middle childhood. He was the eldest of four brothers, all of whom developed conduct problems, but his were by far the most severe. He had a history of becoming involved in aggressive exchanges that often escalated to violence. He and his mother had become involved in coercive patterns of interaction from his earliest years. He developed similar coercive patterns of interaction at school with his teachers, at work with various foremen and also in his relationship with Rita. He had a distant and detached relationship with his father.

Developmental History. From Bill's developmental history, it was clear that he was a difficult-temperament child who did not develop sleeping and feeding routines easily and responded intensely and negatively to new situations. His language development had been delayed and he showed academic difficulties right from his first years in school. On the positive side, Bill had a strong sense of family loyalty to his brothers and parents and did not want to see the family split up.

Psychometric Assessment. From the Child Behaviour Checklists completed by Rita, it was clear that Bill and his three brothers had clinically significant conduct problems and that Bill's were by far the most extreme. A similar pattern emerged from behaviour checklists completed by the boys' teachers. A psychometric evaluation of Bill's abilities and attainments showed that he was of normal intelligence, but his attainments in reading, spelling and arithmetic fell below the 10th percentile. From his subtest profile on the psychometric instruments, it was concluded that the discrepancy between attainment and abilities was accounted for by a specific learning disability.

School Report. The headmaster at the school which Bill and his brothers attended confirmed that Bill had academic, conduct and attainment problems, but was committed to educating the boys and managing their conduct and attendance problems in a constructive way. The headmaster had a reputation (of which he was very proud) for being particularly skilled in managing children with behaviour problems.

Formulation. Bill was an 11-year-old boy with a persistent and broad pattern of conduct problems both within and outside the home. He also had a specific learning disability and peer relationship problems. Factors which predisposed Bill to the development of these problems included a difficult temperament, a developmental language delay, exposure to paternal criminality, maternal depression and a chaotic family environment. The father's incarceration six months prior to the referral led to an intensification of Bill's conduct problems. The conduct problems were maintained at the time of the referral by engagement in coercive patterns of interaction with his mother and teachers; rejection of Bill by peers at school; and isolation of his family by the extended family and the community. Protective factors in the case included the mother's wish to retain custody of the children rather than have them taken into foster care; the children's sense of family loyalty; and the school's commitment to retaining and dealing with the boys rather than excluding them for truancy and misconduct.

Treatment. The treatment plan in this case involved a multisystemic intervention programme. The mother was trained in behavioural parenting skills. A series of meetings between the teacher, the mother and the social worker was convened to develop and implement a plan that ensured regular school attendance. Occasional relief foster care was arranged for Bill and John (the second eldest) to reduce the stress on Rita. Social skills training was provided for Bill to help him deal with peer relationship problems.

Clinical Features

From Table it may be seen that a distinction is made between oppositional defiant disorder, and conduct disorder with the former reflecting a less pervasive disturbance than the latter. In a proportion of cases oppositional defiant disorder is a developmental precursor of conduct disorder (Loeber and Stouthamer-Loeber, 1998). The main behavioural feature of conduct disorder is a pervasive and persistent pattern of antisocial behaviour which extends beyond the family to the school and community; involves serious violations of rules; and is characterized by defiance of authority, aggression, destructiveness, deceitfulness and cruelty. Youngsters with conduct disorder show a limited internalization of social rules and norms and a hostile attributional bias where the youngster interprets ambiguous social situations as threatening and responds with aggressive retaliative behaviour. Anger and irritability are the predominant mood states. Problematic relationships with significant members of the child's network typify children with conduct disorder. Negative relationships with parents and teachers typically revolve around the youngster's defiant behaviour, and with peers the problems typically centre on aggression and bullying which is guided by the hostile attributional bias with which conduct-disordered youngsters construe many of their peer relationships. With conduct disorders there may also be problematic relationships with members of the wider community if theft or vandalism has occurred. Multiagency involvement with juvenile justice or social work agencies is common. Also, because conduct disorder is associated with family disorganization, parental criminality and parental psychological adjustment difficulties, professionals from adult mental health and justice systems may be involved.

With both conduct disorder and oppositional defiant disorder three classes of risk factors increase the probability that conduct problems in childhood or adolescence will escalate into later life difficulties—that is, personal characteristics, parenting practices, and family organization problems (Kazdin, 1995). Difficult temperament, aggressiveness, impulsivity, inattention and educational difficulties are the main personal characteristics of children and adolescents that place them at risk for long-term conduct problems. Ineffective monitoring and supervision of youngsters, providing inconsistent consequences for rule violations, and failing to provide reinforcement for prosocial behaviour are the main problematic parenting practices that place children and adolescents at risk for the development of long-term anti-social behaviour patterns. The family organization problems associated with persistence of conduct problems into adulthood are parental conflict and violence, a high level of life stressors, a low level of social support and parental psychological adjustment problems such as depression or substance abuse.

Epidemiology

Overall prevalence rates for conduct disorder vary from 4 to 14 per cent depending upon the criteria used and the population studied (Cohen *et al.*, 1993). Conservative estimates of prevalence range from 2 to 6 per cent (Kazdin, 1995). Conduct disorder is more prevalent in boys than in girls with male: female ratios varying from 4:1 to 2:1. Comorbidity for conduct problems and other problems, such as ADHD or emotional disorders, is quite common, particularly in clinic populations (McConaughy and Achenbach, 1994). The comorbidity rate for conduct disorder and ADHD in community populations is 23 per cent. The comorbidity rate for conduct disorder and emotional disorders in community populations is 17 per cent for major depression and 15 per cent for anxiety disorders.

Etiological Theories

Biological, psychodynamic, cognitive-behavioural and social systems theories have been put forward to explain the development of conduct problems and to guide the development of treatment programmes. Since the distinction between oppositional defiant disorder and conduct disorder is a relatively recent development, most theories in this area have been developed with specific reference to conduct disorder, but have obvious implications for oppositional defiant disorder which is probably a developmental precursor of conduct disorder in many cases.

Biological Theories

Biological theories have focused on the roles of genetic factors, hormonal factors, arousal levels and neuropsychological deficits in the etiology of conduct problems.

Genetic theories. There are many lines of research which focus on genetic and constitutional aspects of children with conduct disorder, and these are guided by the hypothesis that biological factors underpin antisocial behaviour in youths. The predominance of males among youngsters with conduct disorders and the observation that the concordance rate for conduct problems among monozygotic twins (87 per cent) is higher than that for dizygotic twins (72 per cent) point to the importance of some constitutional factors in the etiology of conduct disorders (Plomin, 1991). Early reports of a causal link between XYY syndrome and conduct problems have not been supported by later research. The association between XYY syndrome and conduct problems appears to be due to the high rates of parental separation and maternal psychological problems that characterize the families of XYY children (Bolton and Holland, 1994).

Hormonal theories. One of the more promising leads in the search for biological factors associated with conduct problems is a link between aggressive behaviour and elevated testosterone levels in male teenagers (Dabbs *et al.*, 1991).

Arousal theory. Children with conduct disorders have lower arousal levels than normal children, according to this theory, and so are less responsive to rewards and punishments (Raine, 1988). They have an impaired capacity for responding to the positive reinforcement that often follows prosocial behaviour or for avoiding punishments associated with antisocial behaviour. Thus, they fail to learn prosocial behaviour or to avoid antisocial behaviour. It is assumed that this abnormally low arousal level is inherited and the results of twin studies partially support this (Kazdin, 1995). Treatment based on this hypothesis must involve highly structured and intensive learning situations if social rules are to be learned. The positive and negative reinforcers used must be highly valued and delivered immediately following responses. All rule infractions must lead to immediate withdrawal of desired stimuli. Rule-following should be immediately and intensely rewarded on a variable interval schedule, since this leads to learning that is maximally resistant to extinction. These treatment implications of arousal theory have been incorporated into the design of residential token economies for delinquent adolescents; behavioural parent training programmes; school-based behavioural programmes; and treatment foster care (Patterson, 1982; Chamberlain, 1994; Patterson *et al.*, 1992).

Neuropsychological deficit theory. Neuropsychologically-based deficits in verbal reasoning and executive functioning, according to this position, underpin self-regulation difficulties that contribute to conduct problems. They may also lead to underachievement which leads to frustration and this contributes to aggressive behaviour. This position is supported by a substantial body of evidence that documents verbal reasoning and executive function deficits in children and teenagers with conduct problems, by studies that confirm a strong association between reading difficulties and conduct problems, and by studies that show that unsocialized conduct problems are associated with self-regulation problems (Moffit, 1993; Shapiro and Hynd, 1995). Remedial interventions that facilitate the development of language and academic skills are the principal types of treatment deriving from this theory.

Psychodynamic Theories

Classical psychoanalytic theory points to superego deficits and object relations theorists highlight the role of disrupted attachments in the development of conduct problems.

Superego deficit theory. Within psychoanalysis it is assumed that societal rules and expectations are internalized through identification with the parent of the same gender. This internalization is referred to as the superego. Aichorn (1935) argued that anti-social behaviour occurs because of impoverished superego functioning. The problems with superego functioning were thought to arise from either overindulgent parenting on the one hand or punitive and neglectful parenting on the other. With overindulgent parenting, the child internalizes lax standards and so feels no guilt when breaking rules or behaving immorally. In such cases any apparently moral behaviour is a manipulative attempt to gratify some desire. With punitive or neglectful parenting, the child splits the experience of the parent into the *good caring parent* and the *bad punitive/neglectful parent* and internalizes both of these aspects of the parent quite separately with little integration. In dealing with parents, peers and authority figures, the child may be guided by either the internalization of the good parent or the internalization of the bad parent.

Typically at any point in time such youngsters can clearly identify those members of their network who fall into the good and bad categories. They behave morally towards those for whom they experience a positive transference and whom they view as good, and immorally to those towards whom they have a negative transference and view as bad. Residential group-based milieu therapy where staff consistently and compassionately enforce rules of conduct which reflect societal standards is the principal treatment to evolve from this theoretical perspective. Within such a treatment programme children gradually internalize societal rules, integrate the *good* and *bad* parental introjects, and develop a more adequate superego.

While there is little evidence for the effectiveness of psycho-analytically-based treatment for conduct disorders (Kazdin, 1995), it has provided important insights into the impact of working with such youngsters on the dynamics within multidisciplinary teams. For example, in my clinical experience, conduct-disordered youngsters who have internalized good and bad parental representations into the superego typically project good parental qualities onto one faction of the multidisciplinary team (typically the least powerful) and bad parental qualities onto the other team members (typically the most powerful). These projections elicit strong countertransference reactions in team members, with those receiving good projections experiencing positive feelings towards the youngster and those receiving bad projections experiencing negative feelings towards the youngster. Inevitably this leads to team conflict which can be destructive to team functioning if not interpreted, understood and worked through.

Attachment theory. Bowlby (1944) pointed out that children who were separated from their primary caretakers for extended periods of time during their first months of life failed to develop secure attachments and so, in later life, did not have internal working models for secure trusting relationships. He referred to such children as displaying affectionless psychopathy. Since moral behaviour is premised on functional internal working models of how to conduct oneself in trusting relationships, such children behave immorally. Treatment according to this position should aim to provide the child with a secure-attachment relationship or corrective emotional experience which will lead to the development of appropriate internal working models. These in turn will provide a basis for moral action.

While the provision of a secure attachment experience within the context of outpatient weekly individual therapy is an ineffective treatment for children with conduct disorder, a secure attachment relationship is a central treatment component in some effective interventions, such as treatment foster care (Chamberlain, 1994). Here, the foster parents provide the child with a secure-attachment experience and couple this with good behavioural management.

Cognitive-behavioural Theories

A range of different theories of conduct problems have been developed within the broad cognitive-behavioural tradition. Problems with social information processing and social skills deficits are the principal factors highlighted in cognitive theories. Social learning theories highlight the importance of modelling, and behavioural theories focus on the role of reinforcement contingencies in the maintenance of conduct problems.

Social information processing theories. The central hypothesis here is that children with conduct disorders process social information in a different way from other children. Research on social information processing in youngsters with conduct disorders has shown that in ambiguous social situations their cognition is characterized by a hostile attributional bias (Crick and Dodge, 1994). Children with conduct disorders attribute hostile intentions to others in social situations where the intentions of others are ambiguous. The aggressive behaviour of children with conduct disorders in such situations is, therefore, intended to be retaliatory. The aggression is viewed as unjustified by those against whom it is directed and this leads to impaired peer relationships. The reactions of peers to such apparently unjustified aggression provide confirmation for the aggressive child that their peers have hostile intentions, which justifies further retaliatory aggression.

Social skills deficit theory. The hypothesis here is that conduct disorders involve inappropriate social behaviours which have developed to compensate for social skills deficits. Research conducted to test this hypothesis has highlighted the social skills deficits of children with conduct disorders (Spivack and Shure, 1982). These children lack the skills to generate alternative solutions to social problems such as dealing with an apparently hostile peer. They also lack the skills to implement solutions to social problems such as these, for example using humour or shared interests to reduce hostility. Within this cognitive-behavioural tradition, group-based social skills programmes have been developed which aim to train youngsters in the social skills they lack. There is a small but growing body of data which shows that group-based social problem-solving skills training is an effective component of broad multisystemic intervention packages for delinquent adolescents (Kazdin, 1995; Chamberlain, 1994).

Modelling theory. Bandura and Walters (1959) have taken the position that aggression, characteristic of children with conduct disorders, is learned through a process of imitation or modelling. In some instances it may be the behaviour displayed by the parents that the child imitates. Fathers of aggressive boys typically are aggressive. Mothers of such children are typically rejecting and discourage the expression of their children's dependency needs. It is this aggression and neglectful hostility that aggressive children are imitating. This position is supported by a large body of evidence, particularly that which points to the intrafamilial transmission of aggressive behaviour (Kazdin, 1995). According to modelling theory, treatment should aim to help parents model appropriate behaviour for their children or provide alternative models of appropriate behaviour in a residential or treatment foster-care setting (Chamberlain, 1994).

Coercive family process theory. This position, which is most clearly articulated by Patter son and his group (Patterson *et al.*, 1992), begins with the hypothesis that children with conduct disorders learn their antisocial behaviours from involvement in coercive patterns of interaction with their parents and these behaviours are then exhibited in school and community contexts. Marital discord, parental psychopathology, a variety of social and economic stressors and social isolation all contribute to the parents' use of a coercive parenting style. This style has three main features. First, parents have few positive interactions with their children. Second, they punish children frequently, inconsistently and ineffectively. Third, the parents of children with conduct problems negatively reinforce antisocial behaviour by confronting or punishing the child briefly and then withdrawing the confrontation or punishment when the child escalates the anti-social behaviour, so that the child learns that escalation leads to parental withdrawal. By middle-childhood children exposed to this parenting style have developed an aggressive relational style which leads to rejection by non-deviant peers.

Such children, who often have comorbid specific learning difficulties, typically develop conflictual relationships with teachers and consequent attainment problems. In adolescence, rejection by non-deviant peers and academic failure make socializing with a deviant delinquent peer group an attractive option. Patterson's group has shown that this developmental trajectory is common among youngsters who first present with oppositional defiant disorder. The delinquency of adolescence is a staging post on the route to adult antisocial personality disorder, criminality, drug abuse and conflictual, violent and unstable marital and parental roles for more than half of all youngsters with conduct disorder (Farrington, 1995). Therapy for families with preadolescent children based on this model aims to help parents and children break coercive patterns of interaction and build positive relationships, but most importantly it helps parents develop skills for effectively disciplining their children. Behavioural parent training and treatment foster care are the principal formats within which such treatment is offered and there is considerable evidence for the effectiveness of both approaches (Kazdin, 1995; Chamberlain, 1994).

Social Systems Theories

Systems theories highlight the role of family systems and broader social systems in the etiology and maintenance of conduct problems. *Family systems theory.* Within the family therapy tradition, the structural, functional and strategic schools have been most influential in offering a framework for understanding how conduct disorders are maintained by patterns of family interaction and how they may be resolved by intervening in these patterns (Colapinto, 1991; Madanes, 1991; Barton and Alexander, 1981). At a structural level, families with youngsters who have conduct problems are more disorganized than other families. Rules, roles and routines are unclear. Communication is indirect, lacking in empathy and confusing. There is also an absence of systematic family problem-solving skills. The members are more emotionally disengaged from each other in comparison with other families. In addition, families with youngsters who display conduct problems have difficulties maintaining clear unambiguous intergenerational hierarchies and negotiating lifecycle transitions.

With respect to ambiguous hierarchies, conduct problems are maintained if a youngster becomes involved in conflicting overt and covert hierarchies with the parents. The overt hierarchy in most families involves a spoken acceptance that both parents share a strong coalition around which there is a boundary which separates the parents hierarchically from the child so that the child is to some degree subservient to the wishes of both parents. Where children have conduct problems, in addition to this overt hierarchy, there is typically a covert hierarchy in which the child and one parent share a strong cross-generational coalition around which there is a covert boundary which separates them hierarchically from the other parent. This covert hierarchy is not spoken about or denied. Usually it is the mother and child who share the covert coalition and the father who is hierarchically inferior to this dyad. Haley (1967) refers to this family structure as the pathological triangle. Not surprisingly, parents in families with this organizational problem often have marital difficulties.

With respect to lifecycle transitions, many families in our culture evolve through a series of predictable stages of the life-cycle. Haley (1980) has argued that some families with youngsters who have conduct problems become stuck at the stage of being a family with adolescents living at home and have difficulty making the transition to being a family in the empty nest stage of development. These families often display the structure described as the pathological triangle. The delinquent behaviour may serve the function of preventing the family from splitting up (Alexander and Parsons, 1982).

Family therapy based upon structural and strategic systems theory aims to help families become more coherently organized. There is some evidence that family therapy based on these structural and functional principles is a more effective treatment for preadolescents and young adolescents with conduct problems than either non-directive family therapy or individual therapy (Alexander and Parsons, 1982; Szapocznik *et al.*, 1989).

Sociological theories. A variety of sociological theories have posited a causal link between deviant antisocial behaviour typical of conduct disorders and aspects of the wider sociocultural context within which such behaviour occurs. Anomie theory is a commonly cited exemplar of this body of theories (Cloward and Ohlin, 1960). According to anomie theory, theft and other related antisocial behaviour such as mugging and lying are illegitimate means used by members of a socially disadvantaged delinquent subculture to achieve material goals valued by mainstream culture. Anomie is the state of lawlessness and normlessness that characterizes such subcultures. Treatment premised on this theory must provide delinquents and their peer groups with legitimate means to achieve societal goals. Remedial academic programmes, vocational training programmes, and treatment foster care are the main treatment approaches implicated by this theory. There is some evidence for the efficacy of treatment foster care (Chamberlain, 1994).

Multisystemic ecological theory. This position entails the view that multiple systems (including the individual, the family, the school and the community) are involved in the genesis and maintenance of conduct problems and consequently effective treatment must target multiple systems rather than any single system (Henggeler *et al.*, 1998). Bronfenbrenner's (1986) model of ecologically nested systems is the foundation for this theory. Conduct disorders, it is argued, are maintained by multiple factors in these multiple ecologically nested systems. Important individual factors include difficult temperament, early separation experiences, hostile attributional bias, poor social skills, difficulties learning prosocial behaviour from experience, and academic learning difficulties. Family factors include family disorganization, ambiguous family hierarchies, parent-child attachment difficulties, parenting and discipline problems, marital discord and difficulty negotiating family lifecycle transitions. School factors include patterns of inter-action that maintain school-based discipline problems, attainment difficulties and lack of educational resources.

Community factors include involvement with deviant peers, drug abuse and involvement in poorly co-ordinated multiagency networks. Treatment based on this model must be individually tailored and based on a multisystem ecological assessment. Treatment packages should include:

- Individual and group cognitive and social skills training
- Family therapy to reduce family disorganization
- School-based interventions to deal with interactional patterns that maintain school-based conduct problems and under-achievement
- Peer-group-based interventions to enhance prosocial peer relationships and reduce involvement in deviant peer groups

There is some evidence that this approach is effective (Henggeler *et al.*, 1998; Brosnan and Carr, 2000).

INTERVENTION

From the foregoing discussion it is clear that conduct problems are complex. Personal characteristics of children and adolescents and psychosocial factors within the family and the wider system may predispose youngsters to develop conduct problems. Once these develop, they may remain confined to the home in the case of oppositional defiant disorder or spread to the community in the case of conduct disorder. The degree to which conduct problems escalate depends upon the degree to which there are problem-maintaining factors within the family, school and peer group. Despite their complexity, conduct problems are not completely refractory to treatment. Available evidence from empirical studies permits treatments of choice to be identified for preadolescent behaviour problems and more pervasive difficulties in adolescence.

For the circumscribed conduct problems, typical of children with oppositional defiant disorder, behavioural parent training is currently the treatment for which there is greatest empirical support (Behan and Carr, 2000; Serketich and Dumas, 1996). Behavioural parent training involves coaching parents to target and reinforce their children's prosocial behaviour while extinguishing their antisocial behaviour.

For pervasive conduct problems, typical of children with conduct disorder, family therapy, multisystemic therapy and treatment foster care are currently the treatments for which there is greatest empirical support (Brosnan and Carr, 2000). Family therapy involves helping parents, adolescents and other family members develop patterns of interaction that promote prosocial behaviour, better communication and family problem-solving. Multisystemic therapy extends beyond the family and includes intervention in the school and peer group and also at the individual level, so that problem-maintaining factors in all of these systems are modified.

Multisystemic intervention may also involve addressing parents' personal difficulties such as depression or alcohol problems. Where parents' personal difficulties are too severe to permit them to engage effectively in multisystemic treatment, treatment foster care is the intervention of choice. Here, the conduct-disordered youngster is placed with a specially trained foster family who work collaboratively with the child's natural parents in implementing a behavioural programme which allows the youngster to develop prosocial behaviour and disengage from patterns of antisocial behaviour and interaction. With respect to service development, it may be most efficient to offer services for adolescent conduct disorder on a continuum of care (Chamberlain and Rosicky, 1995; Carr, 2000a). Less severe cases may be offered family therapy. Moderately severe cases and those that do not respond to circumscribed family interventions may be offered multisystemic therapy. Extremely severe cases and those that are unresponsive to intensive multisystemic therapy may be offered treatment foster care.

There are many other treatments for conduct disorders besides those highlighted in this chapter. Some are empirically untested so we do not know if they are effective. Some have been tested and found to be ineffective. And, most importantly, there are those treatments which have been tested and found to be harmful. There is now considerable evidence that intervention programmes which bring youngsters with conduct disorders together in groups lead to an increase in conduct problems (Dishion *et al.*, 1999). This may occur because youngsters reinforce each other's deviant behaviour and share a commitment to a deviant set of values and ideology. The fact that interventions which aggregate conduct-disordered youngsters are harmful unfortunately has not had a major impact on policies for dealing with juvenile delinquents. Indeed, many of our current policies in this domain exacerbate rather than ameliorate the problem.

CONTROVERSIES

There are many controversies in the scientific study and clinical treatment of abnormal behaviour in childhood. A controversy concerning the diagnosis of conditions such as dyslexia (specific reading disability) or ADHD deserves particular mention because it highlights many of the ethical dilemmas psychologists have to face when studying and treating abnormal behaviour in childhood. There is a view that disorders such as ADHD or dyslexia are invalid fabrications and that giving children these diagnoses meets particular needs of parents, health or educational professionals, the pharmaceutical industry or society (Cowart, 1988; Hutchins and Hind, 1987; Breggin, 1991). For example, parents who have children with literacy problems may prefer to view their youngsters as 'bright but dyslexic' rather than 'slow learners'.

Educational professionals may also support the diagnosis of dyslexia because they have an investment in developing programmes for dyslexic children and having these resourced through state funding. With ADHD, parents or schools may have difficulty meeting children's needs for intellectual stimulation, nurturance and clear limit-setting and so their children become aggressive and disruptive. In response these parents or educational professionals may prefer children in their care to receive a diagnosis of ADHD and a prescription for stimulant therapy, such as Ritalin (methylphenidate), rather than exploring ways to better meet the children's needs for intellectual stimulation, nurturance and clear limit-setting. In such instances pharmaceutical companies may support the diagnosis of ADHD, because they may stand to gain financially from offering a pharmacological treatment for behaviour problems.

There is an important ethical dimension to this position which views diagnoses as invalid fabrications designed to meet the needs of specific constituencies (Kutchins and Kirk, 1999; Newnes *et al.*, 2000). For example, there are serious ethical problems with the diagnosis of a condition which is an invalid fabrication if the treatment for the condition (such as prescribing Ritalin) has harmful effects. Currently, the long-term effects of Ritalin or other stimulant therapies are unknown. Furthermore, if dyslexia and ADHD are invalid fabrications, then it is difficult to ethically justify committing public resources to funding their treatment when these resources might be better used to deal with the more fundamental problems which lead to society wanting to believe in diagnoses such as dyslexia and ADHD. These problems might be the stigmatization of slow learners or the reticence of society in supporting parents and teachers in developing ways to meet their children's needs for intellectual stimulation, nurturance and clear limit-setting.

An alternative to this viewpoint is that there is a large body of scientific evidence which supports the validity of diagnoses such as dyslexia and ADHD. There is evidence which shows that many children with dyslexia score high on IQ tests but poorly on measures of visual and auditory sequential memory (Thomson, 1990). Such children have difficulties processing symbolic information and these difficulties are subserved by abnormal neurophysiological processes. Advocates of this viewpoint accept that dyslexia is a well-defined syndrome which deserves further scientific investigation. They also view the development and resourcing of remedial teaching programmes for dyslexic children as essential and argue that to fail to do this is discriminatory and unethical.

Similarly psychologists who study ADHD have shown that these children have difficulties sustaining attention and regulating their activity levels and these difficulties are subserved by abnormal neurophysiological processes (Barkley, 1998). Further-more, many children with a diagnosis of ADHD come from families in which parents have demonstrated good parenting skills with other siblings, so it may be argued that ADHD is not the outcome of poor parenting. Advocates of this viewpoint accept that ADHD is a well-defined syndrome requiring further scientific investigation and argue that it would be unethical to withhold pharmacological treatments such as Ritalin (or other stimulant therapies) for children with this disability.

In order to further our understanding of abnormal behaviour (such as literacy problems, attention difficulties and overactivity) continued scientific study is essential. In so far as the use of diagnostic categories as working hypotheses supports this, diagnoses such as dyslexia and ADHD are valuable. However, it is also valuable to study the social psychological processes that underpin diagnosis in clinical practice and to be vigilant for ethical problems that this may entail.

Attention deficit hyperactivity disorder is currently the most common term used for a syndrome characterized by persistent overactivity, impulsivity and difficulties in sustaining attention. Between 1 and 5 per cent of children have this syndrome, which is typically lifelong. Comorbid developmental language delays, specific learning difficulties, elimination disorders, conduct disorders and emotional disorders are quite common. A poor outcome occurs for about a third of cases who typically have secondary conduct and academic problems. Available evidence suggests that a marked genetic predisposition for an overactive temperament, which finds expression as a result of exposure to physical and psychosocial environmental risk factors during the pre- and perinatal periods and early infancy, causes the syndrome. Adjustment problems shown by youngsters with ADHD are in part maintained by problematic relationships within the family, school and peer group. Multimodal treatment includes behavioural parent training, school-based contingency management, self-regulation skills training, dietary control where food intolerance is present, and stimulant therapy. In addition assessment and treatment of comorbid problems may be required.

Conduct disorders are the most common type of childhood psychological problems. Children with conduct problems are a treatment priority because the outcome for more than half of these youngsters is very poor in terms of criminality and psychological adjustment. In the long term the cost to society for unsuccessfully treated conduct problems is enormous. Up to 14 per cent of youngsters have significant conduct problems and these difficulties are far more common among boys. Comorbidity for conduct disorders and both ADHD and emotional problems such as anxiety and depression is very high, particularly in clinic populations. The central clinical features are defiance, aggression and destructiveness; anger and irritability; pervasive relationship difficulties within the family, school and peer group; and difficulties with social cognition.

Specifically, there is a failure to internalize social norms and a negative bias in interpreting ambiguous social situations. Biological theories have focused on the roles of genetic factors, hormonal factors and arousal levels in the etiology of conduct problems. Classical psychoanalytic theory points to superego deficits, and object relations theorists highlight the role of disrupted attachments in the development of conduct disorders. Problems with social information processing and social skills deficits are the principal factors highlighted in cognitive theories of conduct problems. Modelling and coercive family processes have been identified by social learning theory as central to the development and maintenance of conduct difficulties. Systems theories highlight the role of characteristics of family systems, broader social network systems and societal systems in the etiology and maintenance of conduct problems. With oppositional defiant disorders in preadolescent children whose problems are confined to the home, behavioural parent training is the treatment of choice. With older children and adolescents who present with pervasive conduct problems, a multisystemic intervention programme targeting specific problem-maintaining processes or potential problem-resolving processes within the child, the family and the school is the most effective approach to treatment.

3

Anxiety Disorders

Everyone feels a little nervous sometimes. However, people with anxiety disorders feel an abnormal amount from common things. In all types of anxiety disorders, (not surprisingly) anxiety is the main symptom. There are four major types of anxiety disorders: generalized anxiety/panic disorders, phobias, obsessive-compulsive disorders, and post traumatic stress disorder.

General Anxiety/Panic Disorders

A person with a general anxiety disorder lives in a state of constant nervousness. People with this disorder usually overreact to any type of stress. Typically, individuals have trouble making decisions and when they actually do this it usually only causes additional worries. Sufferers of general anxiety tend to have panic attacks. Some theorists think that this disorder is caused by a learned anxiety. For example, being nervous once about something and learning to avoid that something.

Phobias

A phobia is a fear of a specific stimulus or situation. The sufferer of a phobia usually knows that the fear is irrational but cannot do anything about it. Phobia has three sub-classes: simple phobia, social phobia, and agoraphobia.

Simple Phobias

A simple phobia is a fear of a specific thing or situation. A person may have one phobia but be normal in all other aspects. However, in serious cases, a person may have multiple phobias that interfere with their everyday life.

Social Phobias

Individuals with this class of phobia have an extreme fear of social situations and of embarassing themselves. The most common types of this phobia are public speaking and eating in public.

Agoraphobia

This is the most common phobia that people seeking professional help have. It is also the most difficult to cure. This type of phobia creates an irrational fear of unfamiliar situations. People with agoraphobia avoid open spaces, crowds, traveling, and in extreme cases do not even leave their home.

Obsessive-Compulsive Disorders

"An obsession is the persistent intrusion of unwelcome thoughts, images, or impulses, that cause anxiety. A compulsion is an irresistable urge to carry out certain acts or rituals that reduce anxiety." These two things are often linked together. Individuals with obsessive-compulsive disorder know that their behavior is irrational, but are unable to resist. The resistance only causes them to become anxious and only the carrying out of the act will relieve that anxiety. Compulsion has many forms, but the two most common are washing and checking. The obsessive-compulsive disorder is related to the phobias in that both cause severe anxiety and a patient may suffer from both disorders.

Post-Traumatic Stress Disorder

Post-traumatic stress disorder is caused by a traumatic event that overwhelms a person and ruins their ability to cope with a situation. It can cause flashbacks, nightmares, insomnia, and/or guilt. It is usually extremely long lasting.

While normal fear is adaptive and prevents people from entering threatening situations, with anxiety disorders people develop irrational fears of situations which do not threaten their survival. They also develop non-adaptive behavioural patterns associated with avoidance of feared situations or experiences. For people with anxiety disorders, their fears are accompanied by intense physiological arousal which is shown by the presence of some of the following features: accelerated heart rate; sweating; trembling; sensations of shortness of breath or smothering; feelings of choking; chest pain; nausea; numbness or tingling; or chills or hot flushes. The person may also experience dizziness; derealization (feelings of unreality); or depersonalization (being detached from the self).

After considering the clinical features, epidemiology and developmental aspects of a number of anxiety disorders, theoretical explanations for anxiety will be presented later in this chapter. Each of these specific explanations has been developed within the context of one of four broad theories. These are the biological, psychodynamic, cognitive-behavioural and family systems theories of psychological problems. Later, these four broad theories are reviewed with reference to their main attributes, their contributions to our understanding and treatment of psychological problems, and their limitations.

Anxiety Disorders

Within DSM IV and ICD 10 distinctions are made between a variety of different anxiety disorders based on the developmental timing of their emergence, the classes of stimuli that elicit the anxiety, the pervasiveness and the topography of the anxiety response, and the role of clearly identifiable factors in the etiology of the anxiety. The following are the principal anxiety disorders described in DSM IV and ICD 10:

- Separation anxiety
- Phobias
- Generalized anxiety disorder
- Panic disorder
- Post-traumatic stress disorder
- Obsessive-compulsive disorder

Descriptions of these conditions and case examples are given below.

Separation Anxiety

With separation anxiety, inappropriate fear is aroused by separation from an attachment figure. Although not the only cause of school refusal, it is one of the most common causes of this complaint. Separation anxiety with chronic school refusal is a serious condition since it has such a poor prognosis if left untreated. As many as a third of youngsters with this condition go on to develop panic disorder with secondary agoraphobia (Tonge, 1994).

Case Example

Barry, aged 11, was referred because he had not attended school for two months, following the Easter holidays in the year prior to his entry to secondary school. The family doctor could find no organic basis for the abdominal pain or headaches of which he periodically complained, particularly on the mornings when his mother asked him how his health was. Barry's friends visited him at weekends and he went off cycling with them regularly. But on Monday mornings he was unable to get to school both because of the abdominal pains and also because of a sense of foreboding that something dangerous might happen to his mother. If forced to go to school he would become tearful or aggressive.

Family History. While there was no serious threat to Barry's mother's health, she had a variety of complaints including rheumatism and epilepsy which compromised her sense of well-being. Her epilepsy was usually well controlled, but she had experienced a number of grand mal fits in the six months prior to the referral. Barry was one of four children and all had histories of school refusal. Barry's three brothers, aged 20, 25 and 30, all lived at home and had few friends or acquaintances. His eldest brother ran a computer software business from his bedroom. All of the boys had very close relationships with their mother and distant relationships with their father. The father, Martin, who was a healthy man, ran a corner shop and worked long hours. He left early in the morning and returned late at night. He was very concerned for the welfare of his son, Barry, and believed that his wife mollycoddled the boy. But he was reluctant to challenge her because he did not want to upset her. The parents had a history of marital discord and over the year prior to the referral had strongly disagreed about how to handle the boy's separation anxiety.

Two of Barry's uncles had psychological adjustment difficulties and both had been on medication, although details of their problems were unavailable. These uncles had lived at home with their mother until her death and both of them and Barry's mother and her sister Gina had very close relationships with their mother, Mary, but distant relationships with their father. Barry's mother's parents had also quarrelled about how best to manage the children, with Mary being lenient and her husband being strict. Thus, the pattern of relationships in both Barry's mother's family of origin and Barry's family was very similar. At school, Barry was very popular, particularly because he generously shared candy and sweets from his father's shop with his peers. He had complained of bullying once or twice and on one occasion said the gym teacher victimized him.

Psychometric assessment showed that Barry was of high average intelligence and his attainments in reading, spelling and arithmetic were consistent with his overall level of ability. His school reports were good and he was in the top third of his class with respect to ability.

Formulation. Predisposing factors in this case include a possible genetic vulnerability to psychological adjustment problems and the modelling experience of seeing his three brothers develop separation anxiety and subsequent school refusal. Barry's anticipation of the transition to secondary school in the autumn and his awareness of his mother's worsening health may have precipitated the onset of the separation anxiety. The anxiety and the school refusal were maintained by his mother's over concern and the father's limited involvement in the management of the child's difficulties. They may also have been maintained by the availability of an active social life within the house (involving contact with his mother and three brothers) during school hours and outside school hours at the weekend. Protective factors included the good relationships that Barry had with both his teachers and peers at school and the willingness of the father to become involved in treatment.

Treatment. Treatment involved a series of family sessions and home-school liaison with the parents and school staff. Martin, the father, eventually agreed to drive Barry to school regularly for a month and for a teacher to meet Barry in the car park and bring him into the classroom where he was to sit with two peers and work on a special project for twenty minutes before class started each day. Concurrently, weekly family sessions were held in which progress was assessed; a reward system for school attendance was set up; and the transition to secondary school was discussed. The mother also arranged for a series of consultations for her epilepsy which became better controlled. Barry returned to school and moved to secondary school in the autumn. His recovery, however, was incomplete and he later relapsed and required further treatment.

Phobias

Phobic anxiety is the intense fear which occurs when faced with an object, event or situation from a clearly definable class of stimuli which is out of proportion to the danger posed by the stimulus, and leads to persistent avoidance. In DSM IV specific phobias are distinguished from social phobias and agoraphobia. Specific phobias are subdivided in DSM IV into those associated with animals, injury (including injections), features of the natural environment (such as heights or thunder), and particular situations (such as elevators or flying). With social phobias the principal fear is of being evaluated by other unfamiliar people and behaving in an embarrassing way while under their scrutiny. Social phobia leads to a constriction of social life. In earlier versions of the DSM the term avoidant disorder has been used to designate this condition.

Case Example

Nora, aged 9, was referred because of her fear of the dark. She wanted to go on a camping trip with the Brownies but was frightened because she would have to sleep in complete darkness. This was something she had never done. She always slept with the light on in her bedroom and with the door open and the landing light on. Her developmental history was unremarkable and she had never experienced a traumatic incident in the darkness. Her parents had tried to persuade her to sleep with the light off, but she became so distressed on these occasions that they had stopped making such attempts and believed that she would eventually grow out of the darkness phobia. Nora was an only child and there was no family history of anxiety disorders or adjustment problems, nor was there a developmental history of a particularly traumatic incident.

Treatment. This uncomplicated simple phobia was treated with *in vivo*, parent-assisted systematic desensitization. That is, with support from her parents, on successive nights Nora was helped to cope with sleeping in an increasingly darker and darker room at home. The level of illumination was decreased by leaving the light outside her room illuminated and gradually closing the door further and further. After a month of this treatment programme, Nora went camping and successfully slept in a dark tent for three nights without experiencing undue anxiety.

Generalized Anxiety Disorder

When individuals experience generalized anxiety, they have an ongoing apprehension that misfortunes of various sorts will occur. Their anxiety is not focused on one particular object or situation. In earlier versions of the DSM the term overanxious disorder has been used to designate this condition.

Case Example

Margie, aged 10, was referred because of excessive tearfulness in school which had been gradually worsening over a number of months. The tearfulness was unpredictable. She would often cry when spoken to by the teacher or while playing with her friends during break-time. In the referral letter her family doctor described her as a *worrier, like her mother.*

Presentation. In the intake interview Margie said that she worried about many routine daily activities and responsibilities. She worried about doing poorly at school, that she had made mistakes which would later be discovered, that her friends wouldn't like her, that her parents would be disappointed with the way she did her household jobs, that she would be either too early or too late for the school bus, that there would be no room for her on the bus and that she would forget her schoolbooks. She worried about her health and had frequent stomach-aches. She also had wider-ranging fears about the safety of her family. She worried that the house would be struck by lightning, that the river would break its banks and flood the low-lying fens where she lived and that her house would be washed away. She had concerns about the future and worried that she would fail her exams;

be unable to find a satisfactory job; fail to find a marital partner or marry an unsuitable person. She reported feeling continually restless and unable to relax.

Family History. Margie was the eldest of four children and the only girl in the family. Both of her parents showed symptoms of anxiety in the intake interview and her mother had been treated with benzodiazepines for anxiety over a number of years. The parents regularly discussed their worries about their own health, safety and their own concerns about the uncertainty of the future. The father, Oliver, worked with an insurance company, and frequently discussed accidents and burglaries that had befallen his clients. Margie regularly participated in these conversations, being the eldest child. The parents'chief concern was about Margie's tearfulness which they viewed as unusual. Her worries and fears they saw as quite legitimate. Margie had a couple of close friends with whom she played at the weekends, but spent a lot of time in her parents' company.

Formulation. Predisposing factors in this case include a possible genetic vulnerability to anxiety and exposure to a family culture marked by a concern with safety and an oversensitivity to danger. No clearcut precipitating factor is apparent. Ongoing involvement in parental conversations about potential threats to the well-being of family members possibly maintained the condition along with the attention and concern shown towards Margie's tearfulness at school.

Treatment. Treatment in this case involved family work focusing on helping Margie and her parents reduce the amount of time they spent talking about danger and threats and increase the amount of time they spent engaged in activities and conversations focusing on Margie's strengths and capabilities. The parents were also helped to coach Margie in relaxation skills and mastery-oriented coping self-statements. Some reduction in anxiety and tearfulness occurred and Margie showed some improvement in her adjustment in school.

Panic Disorder

With panic disorder there are recurrent unexpected panic attacks. These attacks are experienced as acute episodes of intense anxiety and are extremely distressing. The person may experience a fear of losing control; a fear of going crazy; or a fear of dying. People with this condition come to perceive normal fluctuations in autonomic arousal as anxiety-provoking, since they may reflect the onset of a panic attack. Commonly secondary agoraphobia develops. The person fears leaving the safety of the home in case a panic attack occurs in a public setting.

Case Example

Sandra, a 15-year-old girl, was referred because of anxiety about sitting exams. She lived with her grandparents, Ruth and Josh. She slept and ate well and appeared to be happy. However, she would not venture away from the house. A tutor from the local technical college at which she was enrolled had regularly brought school-work to her for about nine months. The imminence of her GCSE exams, which were due to be held at the college, precipitated the referral. She wanted to overcome her anxiety so that she could travel to college and sit her exams, which she had felt unable to complete the previous year due to anxiety.

History of the Presenting Problem. In a preliminary interview, conducted at her grandparents' house, she described a fear of leaving the safety of her own home and how this fear increased with distance from her house. The fear began during her mock GCSE school exams a year previously. She had a panic attack and left the exam hall. She ran to her grandparents' house after this incident and subsequent attempts to return to school led to further panic attacks. Her family and the college staff, after some preliminary ineffective attempts to help her get out and about, gave up trying. On many occasions, when she found herself any distance from the house, she would begin to panic and run back quickly. This led to the symptoms of panic abating. One staff member at the college visited her and taught her some relaxation exercises. He suggested she use these to help her cope with attempts to leave the house, but she found them of little benefit. Eventually she settled for a house-bound life.

During the attacks she was afraid she would die. She couldn't catch her breath and felt dizzy. She also felt as if she were out of her body and as if the world was dreamlike. The attack lasted no more than a few minutes. Subsequent attacks were similar to the first. On a couple of occasions when she had sufficient courage to visit her friends she had panic attacks. She described the fear of taking hot tea which she might not be able to finish without scalding herself, should she experience a panic attack and need to run home quickly. She said she would not like to offend her friends by not finishing her tea. Sandra's principal fears were that she would have an attack and would not get home safely. She was therefore frightened of going on buses or in cars on the motorway. She was frightened also of queuing at the bank.

Family History. Sandra's parents were divorced. Her father, Des, was a police officer in London and had left her mother, Lynn, when Sandra was 7 years old. Lynn lived near the grandparents, in a rural village about a three-hour drive from London. Lynn cohabited with Jeff whom she had met while hospitalized for depression. The mother had an extensive history of psychiatric treatment for anxiety disorders and depression. The grandparents and the mother were preoccupied with physical illness and psychological problems and regularly discussed threats to each other's well-being. There was also a view, based on Lynn's experiences, that psychological problems ran a chronic course and were unresponsive to psychological treatments.

The grandparents and the mother had very close relationships with Sandra. Lynn was involved in regular conflicts with her mother over the suitability of Jeff as a partner for her. Sandra's brother, Paul, who attended university, visited her occasionally with his friends and she envied his lifestyle. He rarely joined in the conversations about illness at the grandparents' house. He was a drama enthusiast and Sandra would help him rehearse his lines when he visited. For Sandra, this was a welcome break from the regular conversation at her grandparents' house. Sandra had four or five friends who lived locally and two of these visited regularly.

Formulation. In this case the principal predisposing factors were a genetic vulnerability to anxiety from the mother's side of the family and a family culture that focused on illness, fear, danger and anxiety. The exam situation was the principal precipitating factor. Multiple unsuccessful treatments and the experience of negative reinforcement afforded by escaping from threatening situations maintained the condition. Other maintaining factors included the father's lack of involvement in attempts to help Sandra recover, the grandparents' and mother's danger-saturated family culture and their beliefs that psychological problems had a chronic course and were unresponsive to treatment. However, two positive peer relationships and a desire for vocational progression were also present in this case and were important protective factors.

Treatment. Treatment in this instance began with family work involving the grandparents, the mother and on a couple of occasions the father, to reduce the amount of illness and anxiety-focused conversation to which Sandra was exposed and to challenge the beliefs that psychological problems were unresponsive to psychological treatments. This was followed with *in vivo* systematic desensitization coupled with a brief trial of clomipramine (Anafranil). *In vivo* systematic desensitization involved Sandra being supported to make increasingly longer outings from her house, while concurrently using relaxation exercises to help her to manage the anxiety these outings evoked. Sandra could not tolerate the side-effects of clomipramine, so the medication was discontinued. It was also arranged for her to sit exams at school in a private room. Following this, work placements at a creche and at an old folks home were arranged by the college staff. While Sandra made a good recovery, she suffered periodic relapses and re-referred herself for a number of further episodes of treatment over the following two years.

Post-traumatic Stress Disorder

Post-traumatic stress disorder (PTSD) occurs in many people following a catastrophic trauma which the individual perceived to be potentially life-threatening for themselves or others. In PTSD there are recurrent intrusive memories of the trauma which lead to intense anxiety. The person attempts to avoid this by suppressing the memories and avoiding situations that remind them of the trauma.

Case Example

Margaret, a 25-year-old woman, was referred because of recurrent nightmares and erratic behaviour at work. She was employed as a cashier in a petrol station in a busy suburban area. On two occasions, while at work, a man armed with a hypodermic syringe filled with blood, which he claimed was HIV-infected, had coerced her into handing over the contents of the cash register.

She subsequently suffered from nightmares and daytime flash-backs. She also suffered from a high level of generalized physiological arousal and was anxious and short-tempered at work. She attempted to deal with the nightmares and flashbacks by putting them out of her mind and thinking about other things, but found that this was becoming less and less effective. When she became flooded with feelings of anxiety she would become inappropriately aggressive to customers at work and was frightened that she would lose her job because of this inappropriate behaviour. At home her relationships with her mother and sisters had deteriorated.

Treatment. Treatment involved Margaret writing down accounts of her dreams and flashbacks and organizing these into a sequence from least to most threatening. She was also invited to alter the endings to these scenarios so that she emerged victorious rather than victimized at the conclusion of each of them. For example, in one scenario, rather than the aggressor successfully attacking her with the HIV-infected syringe, she imagined him shrinking to half his size and then she overpowered him easily. In therapy sessions, Margaret was imaginally exposed to these scenarios, beginning with the least threatening and concluding with the most threatening, until she could vividly imagine each of them without being overwhelmed with anxiety. During the imaginal exposure sessions, the therapist helped her enter a state of deep relaxation and then read out the account of the scenario to her. She coped with the anxiety that this led to by using relaxation and deep breathing exercises in which she had been coached, and also by concluding the scenario by emerging victorious rather than victimized. Her symptoms abated over a six-month period.

Obsessive-compulsive Disorder

Obsessive-compulsive disorder (OCD) is a condition typically characterized by distressing obsessional thoughts or impulses, on the one hand, and compulsive rituals which reduce the anxiety associated with the obsessions, on the other.

Case Example

April, a 35-year-old woman was referred because she had, gradually over a two-year period, developed some unusual habits, beliefs and feelings. With respect to her behaviour, she scrubbed the floors and walls of the kitchen, bathroom and toilet every day and put a full set of bathroom towels in the dustbin after a student lodger had used them once. She also put crockery and dishes she had used to cook the family's dinner in the dustbin. She prevented the children from playing anywhere where they might fall and cut themselves, including the playground. She took an hour to go to bed each night because she had to repeatedly go downstairs to check that the doors were locked and the fire was extinguished. She had been an affectionate person, but now balked if her friends tried to embrace or kiss her.

With respect to her beliefs, she was frightened that she or her children or husband would catch HIV and develop AIDS. She feared that the student lodger or her friends might infect her with the HIV virus. She was also frightened that a burglar might break in if she did not lock up at night and stab her with a needle and infect her or the children with the HIV virus. With respect to her mood, she felt anxious much of the time and had difficulty sleeping. She also felt sad and empty. She was embarrassed, because she knew that her fears were unfounded.

Family history. April was brought up by strict parents with whom she continued to have close contact. She had trained as a nurse but now was a homemaker with a caring and successful husband and two healthy children. She devoted herself fully to the welfare of her children and her husband. Before the onset of her problems, in every way she was an exemplary wife and mother. While on duty as a nurse she pricked her finger with a needle and this led to her first thought of AIDS contamination. Her HIV test was negative, but she could not accept this and developed the obsessional belief that she, her kids and her husband would get AIDS. The belief became stronger when her husband changed job, her younger child went to playschool and she took in a lodger.

Her family responded to her condition in the following way. Her husband helped with her cleaning and checking rituals and reinforced them. Her children did not object to overprotection. Her friends were very understanding of her lack of physical affection. Her sisters discussed her fear of AIDS with her regularly in a sympathetic manner.

Formulation. April was predisposed to developing OCD by two main factors. First, she came from a family where control and cleanliness were valued. Second, she was also highly medically aware of the risk of possible infection with the HIV virus. The OCD was precipitated by the event where she pricked her finger with the hypodermic needle. The OCD was maintained in the following way. She found that her compulsions to clean, discard food, over-protect the children, and check the house over every night relieved her anxiety so she repeated these actions compulsively. Her family and friends reinforced her obsessional thoughts and her husband participated in her compulsive behaviour. There were two protective factors in this case deserving mention. April was very intelligent and able to take on board a formulation of her problem and understand its relevance to treatment. Her husband was prepared to be involved and enlist family help in combating the compulsions.

Treatment. April was treated with a multimodal programme which included antidepressant medication (clomipramine/Anafranil) and a spouse-assisted behaviour therapy programme of exposure and response prevention. She drew up a list of situations in which she developed her obsession, from the least to the most anxiety-provoking. She planned to expose herself to these situations in order of increasing provocativeness and not engage in compulsions while doing so (with her husband's support) until her anxiety abated. For example, she would lie in bed and not check everything while her husband talked reassuringly with her. She responded well to treatment which was carried out over a three-month period.

Clinical Features

The clinical features of the six types of anxiety disorders described above are presented in Table. Clinical features in the domains of perception, cognition, affect, arousal, behaviour and interpersonal adjustment are given. With respect to perception, the six disorders differ in the classes of stimuli which elicit anxiety. With separation anxiety, the stimulus is separation from the caregiver. For phobias it is specific creatures (e.g. animals), events (e.g. injury), or situations (e.g. meeting new people) that elicit anxiety. With generalized anxiety disorder, the person interprets many aspects of their environment as potentially threatening. In panic disorder, somatic sensations of arousal such as tachycardia are perceived as threatening since they are expected to lead to a full-blown panic attack. With PTSD, internal and external cues that remind the person of the trauma that led to the disorder elicit anxiety. With OCD, stimuli that evoke obsessional thoughts elicit anxiety. For example, potentially dirty situations may evoke obsessional ideas about cleanliness and anxiety about contamination.

Cognitions in all six anxiety disorders have the detection and/or avoidance of danger as the central organizing theme. In separation anxiety, children believe that they or their parents will be harmed if separation occurs. With phobias, the person believes that contact with the feared object or creature, or entry into the feared situation, will result in harm, such as being bitten by a dog in the case of dog phobia or being negatively judged by strangers in the case of social phobia. With generalized anxiety, people catastrophize about many features of their environment. For example, they may fear that the house will burn down, their car will crash, they will be punished for wrongdoing, their friends will leave them and so forth. In panic disorder, the person believes that further panic attacks may be fatal and often secondary beliefs evolve that lead to agoraphobia. That is, individuals believe that, provided they stay in the safety of the home, the panic attacks are less likely to occur. With PTSD, there is a belief that, provided the memories of the trauma are excluded from consciousness, the danger of re-experiencing the intense fear and danger associated with the trauma that led to PTSD can be avoided. With OCD, the most common obsessions are with dirt and contamination; catastrophes such as fires, illness or death; symmetry, order and exactness; religious scrupulosity; disgust with bodily wastes or secretions such as urine, stools or saliva; unlucky or lucky numbers; and forbidden sexual thoughts. There is also the belief that engaging in specific rituals will neutralize the threat posed by specific obsession-related stimuli.

Table.

Clinical features of anxiety disorders

	Separation anxiety	*Phobias*	*Generalized anxiety disorder*	*Panic disorder*	*PTSD*	*Obsessice compulsive disorder*
Perception	• Separation is perceived as threatening	• Specific objects, events or situations are perceived as threatening	• The whole environment is perceived as threatening • The child is hypervigilant, scanning the environment for threats to well-being	• The recurrence of a panic attack is seen as threatening • Attention is directed inward and benign somatic sensations are misinterpreted as threatening	• Cues that remind the person of the trauma are perceived as threatening • Hallucinations or illusions may occur where aspects of the trauma are reperceived	• Specific situations, such as those involving dirt, are perceived as threatening and elicit obsessional thoughts
Cognition	• The child believes that harm to the parent or the self will occur following separation	• The child believes that contact with the phobic object or entry into the phobic situation will lead to catastrophe	• The child catastrophizes about many minor daily events	• The youth believes that the panic attacks may lead to death or serious injury	• Recurrent memories of the trauma occur • The child tries to distract him- or herself from recalling these traumatic memories	• Obsessional thoughts, images or impulses intrude into consciousness and may involve themes of contamination, sex or aggression

• The child tries to exclude these thoughts from consciousnes s

(Cont...)

	• The child tries to exclude these thoughts from consciousnes s					
Affect	• Intense fear or anger occurs when separation is anticipated, during separation or following separation	• Intense fear or anger is experienced if contact with the feared object or situation is anticipated or occurs	• A continual moderately high level of fear is experienced, often called free-floating anxiety	• During panic attacks intense fear occurs and between attacks a moderate level of fear of recurrence is experienced	• Against a background of hyperarousal, periodic intrusive episodes of intense fear, horror or anger like those that occurred during the trauma are experienced • The child feels emotionally blunted and cannot experience tender	• The obsessions cause anxiety

(Cont...)

	Separation anxiety	*Phobias*	*Generalized anxiety disorder*	*Panic disorder*	*PTSD*	*Obsessive-compulsive disorder*
Arousal	• Episodes of hyperarousal occur • Sleep problems	• Episodes of hyperarousal occur • Sleep problems	• Continual hyperarousal occurs • Sleep problems	• Episodes of extreme hyperarousal occur against a background of moderate hyperarousal • Sleep problems	• Episodes of extreme hyperarousal occur against a background of moderate hyperarousal • Sleep problems	• Ongoing moderate hyperarousal occurs • Hyperarousal occurs when compulsions are resisted
	• Separation is avoided or			• The person may avoid	• Young children may cling to parents and refuse to	• Motivated by their obsessional beliefs.

(Cont...)

Behaviour	resisted • The child refuses to go to school • The child refuses to sleep alone	• The phobic object or situation is avoided	• As worrying intensifies social activities become restricted	public places in case the panic attacks occur away from the safety of home. This is secondary agoraphobia	sleep alone • Teenagers or adults may use drugs or alcohol to block the intrusive thoughts and emotions	individuals engage in compulsive rituals which they believe will prevent a catastrophe from occurring or undo some
• Suicidal attempts may occur	potentially threatening event which has occurred • These rituals are usually unrealistic					
Interpersonal adjustment	• Peer relationships may deteriorate • Academic performance may deteriorate	• With simple phobias interpersonal problems are confined to phobic situations • Agoraphobia may lead to social isolation	• Peer relationships may deteriorate • Occupational or academic performance may deteriorate	• If agoraphobia develops secondary to the panic attacks, social isolation may result	• Complete social isolation may occur if the trauma was solitary • Where the trauma was shared, the individual may confine interactions to the group who shared the trauma	• Members of the individual's family or social network may become involved in helping the child perform compulsive rituals and inadvertently reinforce them
Note: Features are based on ICD 10 and DSM IV descriptions of anxiety						

In all six of the anxiety disorders listed in Table the beliefs about threat and danger are accompanied by an affective state, characterized by feelings of tension, restlessness and uneasiness. If the individual is compelled to approach the feared stimuli, or in the case of OCD prevented from executing a compulsive ritual, outbursts of anger may occur. For example, children with separation anxiety may have aggressive tantrums if forced to remain at school while their mothers leave. In PTSD, in addition to the affective experiences of uneasiness and tension, an affective experience of emotional blunting, arising from attempts to exclude all affective material from consciousness, may develop.

The patterning of arousal varies depending upon the frequency with which the person comes into contact with feared stimuli. With separation anxiety, hyperarousal occurs only when separation is threatened, and with specific phobias it occurs only in the presence of the feared object. With generalized anxiety disorder, there is a pattern of ongoing continual hyperarousal. With panic disorder and PTSD there is a moderate level of chronic hyperarousal punctuated by brief episodes of extreme hyper-arousal. These occur in panic disorder during panic attacks and in PTSD when memories of the traumatic event intrude into consciousness. With OCD, specific obsession-related cues evoke acute and intense episodes of arousal.

Avoidance behaviours characterize all anxiety disorders. With specific phobias, these may lead to only a moderate constriction in lifestyle. For example, a person may refuse to engage in sports or athletics or to ride a bicycle because of an injury phobia. However, with separation anxiety, generalized anxiety disorder, panic disorder and PTSD, the avoidance behaviour may lead the person to become housebound. With PTSD, alcohol or drug abuse may occur. Alcohol and drugs are used to reduce negative affect and suppress traumatic memories. With OCD, individuals engage in compulsive rituals to reduce anxiety associated with obsessional thoughts. Common compulsions include: washing; repeating an action; checking; removing contaminants; touching; ordering; and collecting.

Interpersonal relationships are affected by different anxiety disorders in different ways. With simple phobias interpersonal difficulties arise only in those situations where the individual refuses to conform or co-operate with routine activities so as to avoid the feared stimuli. For example, a brief episode of marital conflict may occur if one partner refuses to get in an elevator at a shopping mall because of claustrophobia. With separation anxiety, panic disorder, generalized anxiety and PTSD, complete social isolation may occur. With youngsters, peer relationships and school attendance may cease. With adults, non-attendance at work may occur. With OCD, family members may attempt to reduce the sufferer's anxiety by participating in compulsive rituals or in other instances they may increase anxiety by punishing the individual for his or her compulsive behaviour.

Epidemiology

The lifetime prevalence rate for all anxiety disorders is between 10 and 14 per cent (APA, 1994). The prevalence of separation anxiety in children is 4 per cent. The lifetime prevalence rate of simple phobias is 10-11.3 per cent; for social phobias it is 3-13 per cent; for generalized anxiety disorders it is 5 per cent; for panic disorder it is 1.5-3.5 per cent; for PTSD it is 1-14 per cent; and for OCD the rate is 2.5 per cent (APA, 1994). There are clear gender differences in the prevalence of anxiety disorders in children and adults, with more females than males suffering from anxiety disorders. The main exception to this finding is that equal numbers of males and females suffer from OCD.

Comorbidity among anxiety disorders is quite high, and many people suffer from more than one anxiety disorder (APA, 1994). Anxiety disorders may also occur comorbidly with mood disorders (particularly depression) and substance use disorders. Where substance abuse occurs, often alcohol or drugs are used in a self-medicating way, to help the person manage the anxiety disorder. Specific anxiety disorders are associated with specific personality disorders. A proportion of people with social phobia also have an avoidant personality disorder and a proportion of people with OCD also have an obsessive-compulsive personality disorder. There is a strong link between OCD and tic disorders, particularly Tourette syndrome, a condition characterized by multiple motor and vocal tics, with a proportion of people suffering from both conditions. OCD also occurs in a significant proportion of people with eating disorders such as anorexia nervosa. In children, anxiety disorders may occur comorbidly with conduct disorder, attention deficit hyperactivity disorder and major depression.

Developmental Issues

The development of anxiety disorders may be understood within the context of the development of normal fears. Fear occurs in response to threat and includes cognitive, affective, physiological, behavioural and relational aspects (Strongman, 1996). At a cognitive level, an anxiety-provoking situation is construed as threatening or dangerous. At an affective level, there are feelings of apprehension and tension. At a physiological level, autonomic arousal occurs so as to prepare the person for confronting or escaping from the threatening situation. With respect to behaviour, the individual may either confront the danger or avoid it. The interpretation of situations as threatening and the patterning of approach or avoidant behaviour are determined by the social context within which they occur and the way the individual's response affects members of his or her family and social network.

Development of Fear

Throughout childhood and adolescence the types of stimuli which elicit fear change and these changes parallel developments in cognitive and social competencies and concerns (Ollendick *et al.*, 1994; Klein, 1994; Morris and

Kratochwill, 1991). Stimuli that elicit fear at different stages of development are listed in Table below. In the first six months extreme stimulation such as loud sounds or loss of support elicits fear. However, with the development of object-constancy and cause-and-effect schemas during the latter half of the first year, the child can understand the difference between familiar family members and unfamiliar strangers; so the potential for separation anxiety appears.

Table.

Development of fears and anxiety disorders

Age	*Psychological and social competencies and concerns relevant to development of fears, phobias and anxiety*	*Principal sources of fear*	*Principal anxiety disorders*
Early infancy 0-6 months	• Sensory abilities dominate infant's adaptation	• Intense sensory stimuli	
		• Loss of support	
		• Loud noises	
Late infancy 6-12 months	• Sensori-motor schemas	• Strangers	
	• Cause-and-effect object constancy	• Separation	
Toddler years 2-4 years	• Pre-operational thinking	• Imaginary creatures	• Separation anxiety*
	• Capacity to imagine but		

(Cont...)

	inability to distinguish fantasy from reality	• Potential burglars	
Early childhood 5-7 years	• Concrete operational thinking • Capacity to think in concrete logical terms	• The dark • Natural disasters (fire, floods, thunder) Injury • Animals • Media-based fears	• Animal phobia • Blood phobia
Middle childhood 8-1 1 years	• Esteem centres on academic and athletic performance in school • Rigid orderly routines and rituals of early childhood are supplanted by orderly hobbies such as collecting	• Poor academic and athletic performance	• Test anxiety • School phobia
Adolescence 12-18 years	• Formal operational thought • Capacity to anticipate future dangers • Esteem is derived from peer relationships	• Peer rejection	• Social phobias • Agoraphobia • Panic disorder

* Separation anxiety appears in early childhood but peaks in late childhood

The child fears contact with strangers and separation from caretakers. In early childhood, during the pre-operational period, as the skills required for make-believe and imagination develop, but those for distinguishing fantasy from reality are not yet acquired, the child comes to fear imaginary or supernatural creatures. At this time children's mobility also increases and they come to fear animals and potential burglars. In middle-childhood, as their awareness of the natural world and of the world portrayed in the media develops, they come to fear natural disasters such as floods or thunder and lightning and to experience media-based fears such as epidemics of diseases. In late childhood, failure in academic and athletic performance at school becomes a source of fear. With the onset of adolescence, the period of formal operational thinking, the capacity for abstract thought emerges. The youngster can project what will happen in the future and anticipate with considerable sophistication potential hazards, threats and dangers in many domains, particularly that of social relationships. Fears about peer rejection emerge at this stage. All of these types of fears may persist into adulthood to a greater or lesser degree.

Development of Anxiety Disorders

From Table above it may be seen that the emergence of anxiety disorders follows a developmental course which parallels that of normal fears (Silverman and Rabian, 1994; Klein, 1994). Separation anxiety may present as a clinically significant problem at the transition to pre-school or primary school, although it is noteworthy that separation anxiety disorder is most prevalent among children in late childhood. The onset of animal phobias is most prevalent in early childhood. The onset of test anxiety and other types of performance anxiety peaks in later childhood. Social anxiety, panic disorder, and agoraphobia, which often occurs secondary to panic disorder, tend to first appear in adolescence along with generalized anxiety. Obsessive-compulsive disorder is distinct from the normal rituals of childhood which are prominent in the pre-school years and wane by the age of 8 or 9 years, when hobbies involving collecting and ordering selected objects, toys and trinkets probably take their place. It is often at this time that obsessive-compulsive disorder emerges, although it may also appear in adulthood. From a clinical perspective, typically people are referred for treatment of an anxiety problem when it prevents them from completing developmentally appropriate tasks such as going to school, socializing with friends or managing the demands of the world of work.

Etiological Theories

Theoretical explanations for anxiety disorders and related treatments have been developed within biological, psychoanalytic, cognitive-behavioural, and family systems traditions.

Biological Theories

Biological theories of the anxiety disorders point to the role of genetic factors in rendering people vulnerable to their development, and to dysregulations of particular neurotransmitter systems in particular parts of the brain as central to the etiology of anxiety disorders.

The genetic hypothesis. This hypothesis entails the view that anxiety disorders develop where a person with an inherited vulnerability to anxiety is exposed to threatening environmental stimuli at critical developmental stages when they are primed or prepared to develop fears (Torgersen, 1990). There is an ethological implication here, that sensitivity to particular classes of stimuli at particular developmental stages has an adaptive function, and that this has evolved to protect the survival of the species (de Silva *et al.*, 1977). The genetic hypothesis also entails the view that a dysfunctional biological factor which underpins the process of detecting danger is genetically transmitted in families where anxiety disorders occur. In support of the genetic hypothesis there is a substantial body of evidence which shows that there is a high rate of anxiety disorders in the first-degree relatives of people with such conditions (Torgersen, 1990). However, evidence from more convincing adoptive studies is sparse, and it is clear that the vulnerability to some anxiety disorders, particularly generalized anxiety disorder, is less significant than for others, such as panic disorder.

There is also the evidence mentioned earlier that vulnerability to developing animal phobias and separation anxiety is highest during childhood, whereas social phobias, generalized anxiety disorder and panic disorder with agoraphobia more commonly emerge in adolescence (Klein, 1994). The genetic hypothesis applies to all of the anxiety disorders although it probably has limited relevance for PTSD. Implicit in the genetic hypothesis is the view that pharmacological or psychological interventions should aim to help the person cope with a chronic lifelong disorder. Research efforts should focus on identifying the precise biological factor associated with deficits in threat perception that are genetically transmitted, and the mechanisms of transmission.

The GABA hypothesis. This formulation has been developed specifically to explain phobic and generalized anxiety (Mohler and Okada, 1977; Bernstein, 1994). For both of these conditions it is argued that anxiety develops because of a dysfunction in neurones that normally produce gamma aminobutyric acid (GABA) automatically once a high arousal level has been reached in the central nervous system. GABA is usually released once arousal reaches a certain level. It then binds with GABA receptors on excited neurones which underpin the experience of anxiety. This binding process causes inhibition, a reduction in arousal, and a decrease in experienced anxiety. Treatment of phobias and generalized anxiety with benzodiazepines such as diazepam (Valium) or chlordiazepoxide (Librium) reduces anxiety because these drugs bind to the GABA neuroreceptors with a consequent reduction in arousal and the experience of anxiety. While there is some evidence for the anxiety-reducing effects of benzodiazepines with adults, these effects are usually lost as soon as medication ceases and long-term use of benzodiazepines leads to tolerance and dependence (Roy-Byrne and Cowley, 1998). With children, few controlled trials of benzodiazepines have been conducted and in those that have been conducted benzodiazepines have led to only marginally better relief than placebos (Taylor, 1994b).

The adrenergic-noradrenergic hypothesis. This formulation has been developed specifically to explain the occurrence of panic attacks. Panic attacks are thought to be caused by a dysregulation of adrenergic and noradrenergic subsystems of the autonomic nervous system, particularly in the locus coeruleus. Studies of biochemical indices of increased adrenergic functioning and pharmacological challenge with agents known to affect adrenergic functioning such as lactate acid and C02 support this hypothesis (Klein, 1994). Panic attacks may be induced by ingestion of lactate or inhalation of C02 in adults with panic disorder. Challenge tests may be an aid to diagnosis of panic disorder and C02 inhalation has been used to facilitate exposure treatment with adults, where perception of signs of autonomic arousal have been identified as the feared stimuli (Côté and Barlow, 1993). Tricyclic antidepressants such as imipramine which alter noradrenalin functioning have been used to treat panic attacks. Evidence from a number of controlled trials with adults show that antidepressants may be temporarily as effective as behaviour therapy in the treatment of panic disorder; and with children antidepressants may be a useful adjunct to the psychological treatment of school refusal (Côté and Barlow, 1993; Barlow *et al.*, 1998; Taylor, 1994b; Bernstein, 1994).

Basal ganglia hypothesis. This hypothesis, in which a dysregulation of neural systems within the basal ganglia is thought to cause anxiety, has been developed specifically to explain OCD. In support of the basal ganglia hypothesis, there is considerable evidence that a strong association exists between basal ganglia disease and OCD. For example, there is a very high incidence of OCD in cases of Sydenham's chorea, Huntington's chorea, Wilson's disease, Parkinson's disease and Tourette syndrome (Wise and Rapoport, 1989). Furthermore, damage to the basal ganglia resulting from head injury may precipitate an episode of OCD (Max *et al.*, 1994). Also, surgery that disconnects the basal ganglia from the frontal cortex alleviates severe OCD (Rapoport *et al.*, 1994). Some researchers have speculated that in cases of OCD relatively fixed adaptive behaviour patterns of grooming or self-protection are inappropriately released and these patterns are probably encoded in the basal ganglia (e.g. Swedo, 1989).

Serotonin hypothesis. This hypothesis has been developed specifically to explain OCD, and it is argued that a dysregulation of the serotonergic system underpins OCD. People with OCD are thought to have depleted levels of the neurotransmitter serotonin. In support of the serotonergic hypothesis, both child and adult cases of OCD have been shown to respond to specific serotonergic reuptake inhibitors such as clomipramine (Insel *et al.*, 1985; Leonard *et al.*, 1994; Rauch and Jenike, 1998).

Psychoanalytic Theories

In anxiety disorders, according to classical psychoanalytic theory, defence mechanisms are used to keep unacceptable sexual or aggressive impulses and moral anxiety about their expression from entering consciousness. The unacceptable impulses and related moral anxiety become transformed into neurotic anxiety and expressed as an anxiety disorder. In phobias, the unacceptable impulse is repressed and the neurotic anxiety into which it is transformed is displaced onto a substitute object which symbolizes the original object about which the unacceptable impulses were felt. The key defence mechanism is *displacement*. Thus when people say that they are frightened of a particular object or situation, the psychoanalytic hypothesis is that they are frightened about something else, but have displaced their fear from the original taboo object or event onto a more socially acceptable target. In generalized anxiety disorders, the defences break down and the person becomes overwhelmed with anxiety as the unacceptable impulses continually intrude into consciousness and seek expression. In generalized anxiety disorder, anxiety about a taboo object is displaced onto every available target (Malan, 1979; Bateman and Holmes, 1995).

In Freud's original statement of this hypothesis, in the Little Hans case, where the boy had a horse-phobia, he argued that the taboo fears were castration anxiety, and fear of the father was displaced onto horses (Freud, 1909a). In psychoanalytic treatment, the aim is to interpret the defence, the hidden and forbidden feelings which are being repressed and the associated neurotic anxiety. Malan (1979) refers to these three elements (the defence, the hidden feeling, and the associated anxiety) as the triangle of conflict. During treatment the psychologist draws attention to the parallels between the way in which the person manages the current relationship with the therapist; the past relationship with the parents; and current relationships with other significant people in their lives such as their peers, work colleagues or partners. Malan (1979) refers to these three sets of relationships, which are at the heart of transference interpretations, as the triangle of person. Interpretations of the triangle of conflict and the triangle of person are offered tentatively, at a stage in the therapy when a strong working alliance has been established, and within the context of a coherent psycho-dynamic case formulation.

The idea of displacement is a clinically useful concept when working with anxious patients. In my clinical experience people worried about one thing may say that they are worried about another. However, there is no evidence to support the idea that all anxiety disorders represent displacement of anxiety associated with psychosexual developmental conflicts. There is some limited evidence that psychodynamic therapy is effective with adults and children who have anxiety disorders (Roth and Fonagy, 1996; Fonagy and Moran, 1990).

From a psychoanalytic perspective OCD is explained as the sequelae of toilet-training battles (Freud, 1909b). According to psychoanalytic theory, during the anal phase of development children become angry with their parents' insistence that they use the toilet in an appropriate way. Attempts to express these aggressive impulses are met with sanctions from the parents and so the aggression is repressed. In later life these repressed sexual-aggressive impulses attempt to find expression but this causes anxiety. The aggressive impulses and thoughts are *displaced and substituted* by less unacceptable thoughts or impulses. When these intrude into consciousness, they are experienced as ego-alien because they have been disowned or *isolated*. The anxiety is managed by carrying out a compulsive ritual to *undo* or cancel out the undesirable impulse.

Surveys of people with OCD have invariably failed to identify a higher rate of such parent-child conflicts concerning toilet training in comparison with controls (Milby and Weber, 1991), and there is little evidence that people with OCD can benefit from interpretative psychoanalytic psychotherapy. However, the notion that compulsions reflect the use of the defence mechanism of undoing fits with both clinical observations and with the cognitive-behavioural explanation of OCD.

Cognitive-behavioural Theories

Cognitive-behavioural theories of anxiety disorders point to the importance of learning experiences in their development.

Behavioural theory. The most elaborate conditioning model for phobias is Eysenck's (1979) incubation theory. Eysenck argued that anxiety initially develops through one-trial classical conditioning where a neutral object for which a person is biologically prepared to develop an extinction-resistant fear (CS) is paired with a strongly feared object (UCS) and this object elicits anxiety (UCR). Subsequent exposure to the previously neutral object (CS) elicits mild anxiety (CR). Repeated brief exposure to this previously neutral object (CS) leads to an increase in fear through a process of incubation. Brief exposure leads to anxiety (CR) which is effectively paired with the feared object and on the next brief exposure even more fear is elicited. Exposures remain brief, because the feared situation is avoided, and such avoidance behaviour, through a process of operant conditioning, is negatively reinforced, that is, it leads to the cessation of a noxious stimulus. Incubation is a positive feedback process where, through classical conditioning, fear itself reinforces fear of the phobic object. This whole process occurs outside of cognitive control.

Eysenck also argued that people who are constitutionally more neurotic and introverted are more likely to develop phobic anxiety through this incubation process. He also acknowledged, following Seligman, that phobias develop only to a limited group of stimuli and that humans have preparedness, as a result of evolutionary processes, to develop phobias to these through a process of classical conditioning (de Silva *et al.*, 1977). Treatment for phobias according to this theory involves gradual or complete exposure to the phobic object or an image of it until the anxiety subsides. Thus both *in vivo* (or real life) and imaginal desensitization or flooding procedures may be used to treat phobias, although the theory predicts that *in vivo* methods should be more effective. The use of relaxation skills or other coping strategies is only important in so far as they help the person tolerate remaining in the presence of the feared object. There is still controversy over details of the incubation theory as an explanation for the etiology of phobias. However, a large body of research shows that exposure techniques such as desensitization and flooding, particularly *in vivo*, are very effective methods for treating specific phobias in children and adults (Kendall and Treadwell, 1996; Giles, 1993; Dwivedi and Varma, 1997; Barlow *et al.*, 1998).

Cognitive theory. According to the theory that underpins Beck's cognitive approach to therapy for anxiety disorders, anxiety occurs when life events involving threat reactivate threat-oriented cognitive schemas formed early in childhood during a threatening and stressful experience (Beck *et al.*, 1985). These threat-oriented schemas contain assumptions about the dangerous nature of the environment or the person's health such as: *The world is dangerous, so I must continually be on guard*; or *My health is ailing so any uncomfortable somatic sensation must reflect serious ill health.* They also entail cognitive distortions such as minimizing safety-related events and maximizing threat-related negative events. Individual or group cognitive therapy aims to help clients to challenge their assumptions about the dangerousness of the situations in which they feel anxiety. This involves them learning how to monitor their cognitions and anxiety levels and to test out their cognitions by collecting information and engaging in experiences that allow the validity of their assumptions to be checked. Beck's theory is supported by evidence which shows that anxiety is associated with a threat-sensitive cognitive style and also by the results of treatment outcome studies with adults and children which support the efficacy of cognitive-behavioural approaches to treatment (Williams *et al.*, 1992; Hollon and Beck, 1994; Kendall and Treadwell, 1996; Barlow *et al.*, 1998; Keane, 1998).

Cognitive-behavioural theory of OCD. Cognitive-behavioural explanations of OCD argue that specific environmental stimuli trigger obsessional thoughts and compulsive rituals are used to neutralize these intrusive obsessions. These environmental stimuli come to elicit anxiety through an initial process of classical conditioning. Historically, it is argued they were once paired with anxiety-provoking stimuli, and so became conditioned stimuli. According to Rachman and Hodgson (1987) some people have a particular vulnerability to developing intrusive unacceptable obsessive thoughts. A variety of factors contribute to this vulnerability, including genetically determined hyperarousability; depressed mood; socialization experiences that have led to the development of high moral standards; and a belief system involving specific convictions about the relationships between thought and action, control and responsibility. People who are vulnerable to OCD believe that intrusive and morally unacceptable obsessional thoughts will automatically translate into immoral action; that one should be able to control intrusive immoral obsessive thoughts and impulses; and that one is reprehensible for having thoughts that may lead to actions that would harm others. According to cognitive-behavioural theories, episodes of OCD are usually precipitated by stressful life events, illness or family disruption.

The cognitive-behavioural treatment of OCD is usually based upon assessment of the stimuli which elicit obsessions; the anxiety associated with these; and the compulsive rituals which reduce the anxiety. Treatment typically involves exposure to those cues that elicit obsessional thoughts, and response prevention, where the person is prevented from carrying out the anxiety-reducing compulsive rituals. Substantial evidence exists for the vulnerability factors, precipitating factors and self-reinforcing nature of the vicious cycles of obsessions and compulsions that typify OCD (Rachman and Hodgson, 1987). In addition, there is some evidence that exposure and response prevention are a particularly effective treatment for adults and children with OCD (Franklin and Foa, 1998; March and Mulle, 1996).

Family Systems Theory

There is no well-articulated, integrative theory about the role of the family and the socialization process in the etiology of anxiety. However, a number of hypotheses from diverse sources may be drawn together to offer such an explanation (Perlmutter, 1996; Bloch *et al.*, 1994; Bolton, 1996; Dadds *et al.*, 1992; Blagg, 1987; Combrinck-Graham, 1986). Individuals probably develop anxiety problems when they are socialized in families where significant family members (particularly primary caretakers) elicit, model and reinforce anxiety-related beliefs and behaviours. Furthermore, family lifecycle transitions and stressful life events within the family may precipitate the onset of clinically significant anxiety problems. These problems are maintained by patterns of family interaction that reinforce the individual's anxiety-related beliefs and avoidance behaviour.

Family belief systems that promote anxiety may involve the view that ambiguous situations should be routinely interpreted as threatening or dangerous; that the future will probably entail many hazards, catastrophes and dangers; that inconsequential events in the past will probably reap dangerous, threatening consequences at some unexpected point in the future; that fluctuations in autonomic arousal should be interpreted as the onset of fullblown anxiety attacks; that minor ailments are reflective of inevitable serious illness; and that testing out the validity of any of these beliefs will inevitably lead to more negative consequences than continuing to assume that they are true. Through observing significant family members articulate these beliefs and engaging in family interactions premised on these beliefs, individuals may come to internalize them and develop a personal danger-saturated belief system.

When individuals observe significant family members cope with perceived threats by avoiding rather than confronting the perceived danger, they may adopt this coping strategy themselves. Such anxiety-related beliefs and avoidant coping styles may be inadvertently reinforced when family members acknowledge their validity and do not challenge them. Family lifecycle transitions, such as starting school, moving house, birth of a sibling, and family stresses such as illness, may precipitate the onset of a serious anxiety problem. In such situations, the individual interprets the transition or stress as a major threat and copes by engaging in avoidant behaviour.

Members of the nuclear and extended family may all inadvertently maintain an individual's anxiety-related beliefs and avoidance behaviour by sympathizing with the irrational fears, accepting the individual's danger-saturated view of the situation, and condoning the avoidance behaviour as a legitimate coping strategy. The other family members' own danger-saturated belief systems and personal adjustment problems, if such are present, may prevent them from providing the individual with opportunities to develop the skills required to confront and master feared situations. So, for example, in families where there are marital problems, parental depression, parental alcohol abuse or some other difficulty, the parents may avoid facing these difficulties and focus their attention instead on reassuring an anxious child or arranging extensive medical investigations for anxiety-related somatic complaints. The patterns of family interaction that evolve in such situations may inadvertently maintain the child's anxiety and reinforce the parents' avoidance of their own marital or personal difficulties. Commonly the family members are not consciously aware of the secondary gains associated with these problem-maintaining patterns of interaction. Where adults develop anxiety disorders, their partners may become involved in rituals that help their partners avoid feared situations, and this process may allow the couple to avoid dealing with unresolved marital conflicts over issues such as the distribution of power within the marriage.

Family therapy for children with anxiety disorders aims to support the parents and child in creating opportunities within which the child can develop the skills required to confront and master feared situations. Marital therapy for couples in which one partner has an anxiety disorder involves helping the couple to work as a team and jointly enter situations which are increasingly anxiety-provoking for the person with the anxiety disorder and remain in these until the anxiety subsides. The non-symptomatic partner's role in such programmes is to provide support and encourage the symptomatic partner to engage in active coping strategies.

There is substantial evidence that the parents of a majority of children with anxiety disorders have anxiety disorders themselves and that other types of psychopathology, including depression, are over-represented in the families of youngsters with anxiety disorders (Klein, 1994; Bolton, 1996). There is also considerable evidence that modelling and parenting style play an important role in the transmission of anxiety patterns from parents to children in at least some cases (Silverman *et al.*, 1988; McFarlane, 1987). An increased incidence of stressful life events and marital discord is associated with onset of anxiety disorders in some but not all children (Bolton, 1996; Klein, 1994). There is a small and growing body of research which points to the effectiveness of marital and family interventions in the treatment of anxiety disorders (Estrada and Pinsof, 1995; Dadds *et al.*, 1992; Blagg, 1987; Baucom *et al.*, 1998).

An Integrative Approach

In practice many clinicians take an integrative approach to the assessment and treatment of anxiety disorders. Through careful clinical interviewing of individuals and members of their family, the symptoms of anxiety and the situations in which they occur may be identified and a diagnosis given in accordance with the criteria outlined in ICD 10 and DSM IV. In addition a wide variety of self-report scales, rating scales and structured clinical interviews have been developed to aid the reliable and valid assessment of anxiety disorders in children (Carr, 1999) and adults (Lindsay, 1994).

During the process of assessment the individual and family members may also be invited to keep daily records of fluctuations in feelings of anxiety, related thoughts, related avoidant behaviour and the circumstances surrounding these fluctuations. In developing a treatment programme, available evidence suggests that both psychological and psychophar macological treatments can be effective in treating anxiety disorders and in some instances a combined approach is optimal (Nathan and Gorman, 1998). For separation anxiety, the treatment of choice is a brief family-based behavioural approach like that outlined in the case example for separation anxiety disorder (Blagg, 1987). For simple phobias, treatments that involve gradual or sudden prolonged exposure to the feared situation are the most effective. Benzo-diazepines may temporarily alleviate phobic anxiety but the effects of this treatment are lost when medication is ceased. For panic disorder, with agoraphobia, gradual exposure to increasingly threatening situations, combined with family or therapist support, coping skills training combined with clomipramine or a similar SRI, is the current treatment of choice.

For generalized anxiety disorder, cognitive-behaviour therapy, which involves challenging threatoriented thoughts and learning coping and relaxation skills, is the most effective treatment. Benzodiazepines may temporarily alleviate generalized anxiety but the effects of this treatment are lost when medication is ceased. For OCD, family or individually-based exposure and response prevention programmes combined with clomipramine or similar SRI are the treatment of choice. For PTSD, exposure therapies, which involve recalling traumatic memories and tolerating the anxiety these induce until the anxiety subsides, have been found to be particularly effective. A range of medications including tricyclic antidepressants and SRIs have been found to be an important adjunct to psychological exposure therapies.

The discovery that exposure therapies effectively alleviate anxiety is one of the most important contributions that psychologists have made to the treatment of anxiety disorders. Another important finding is that briefly exposing people to threatening stimuli sensitizes them to these stimuli and increases their anxiety. Thus, non-directive permissive approaches to counselling people with anxiety disorders may actually exacerbate rather than alleviate their anxiety. Many of us who work clinically with people who suffer from anxiety come across people whose condition has worsened as a result of participation in well-intentioned, non-directive counselling.

Controversies

There are many controversies in the scientific study and clinical treatment of anxiety disorders. The medicalization of fear and courage is one deserving particular mention (Breggin, 1991). Within ICD 10 and DSM IV, anxiety disorders are framed as medical conditions requiring treatment, and in practice in many instances pharmacological treatment is favoured because it is cheaper and easier to offer than non-pharmacological alternatives. An alternative viewpoint is that this way of conceptualizing fundamental human experiences such as fear and courage further disempowers people who are already feeling frightened and powerless (Newnes and MacLachlan, 1996). For example, if a person has repeated panic attacks and develops a constricted lifestyle because they are afraid of having a panic attack while away from the safety of their home, it may lead them to believe that they are truly powerless to control their fear if their fear is defined as an illness requiring pharmacological treatment.

A further aspect of this argument is that for many years addictive pharmacological treatments, such as diazepam (Valium) or other benzodiazepines, were routinely prescribed for anxiety disorders. Those who are critical of the medicalization of experiences such as fear and courage would argue that, if a person can understand that panic attacks develop from the misinterpretation of bodily sensations and hyperventilation, then they may use this knowledge and their own courage to take control of their fear. In this way they are empowered to be courageous rather than disempowered by being defined as ill. Those who are critical of the medicalization of distress would argue, in the same way, that a person given a diagnosis of PTSD and prescribed medication to manage the recurrent traumatic memories may also become disempowered. They may develop a belief that they are powerless to control recurrent traumatic memories of experiences such as road traffic accidents, assault with a deadly weapon, or involvement in war or combat. In contrast, if they are helped to understand that traumatic memories of life-threatening events must be repeatedly recalled, processed and integrated into people's overall views of themselves, then this opens up a range of non-pharmacological procedures which trauma victims may follow to help them take control of recurrent, intrusive distressing memories.

In order to further our understanding of apparently irrational fears, post-traumatic distress and courage, continued scientific study is essential. In so far as the use of diagnostic categories as working hypotheses supports this, diagnoses such as panic disorder or PTSD are valuable. However, it is also valuable to study fear, distress and courage as normal psychological processes. It may be fruitful too to study the social processes which underpin the medicalization and the medical treatment of fear in clinical practice and to explore the degree to which this, and alternative conceptualizations of fear, empower clients to be courageous.

Normal fear is an adaptive response to potential threats to safety, while anxiety is a similar non-adaptive response to situations which are not threatening. In DSM IV and ICD 10 a number of anxiety disorders are defined, including separation anxiety, phobias, generalized anxiety disorder, panic disorder (with and without agoraphobia), post-traumatic stress disorder and obsessive-compulsive disorder.

At a clinical level anxiety disorders involve selective attention to potential threats; threat-oriented cognition; physiological arousal; avoidance behaviour; and the disruption of interpersonal relationships so that the individual's lifestyle becomes constricted. The lifetime prevalence rate for all anxiety disorders is between 10 and 14 per cent. The emergence of anxiety disorders follows a developmental course which parallels that of normal fears, with separation anxiety and simple phobias appearing in early childhood, while social anxiety, panic disorder, and agoraphobia—which often occurs secondary to panic disorder—have their onset in adolescence, along with generalized anxiety. Biological theories of anxiety implicate dysregulations of a number of systems—notably the GABA system, the adrenergic-noradrenergic system and the serotonergic system—in the etiology of anxiety. Benzodiazepines and serotonin reuptake inhibitors have been shown to partially rectify these difficulties and lead to symptomatic relief. Psychoanalytic theories implicate the defence mechanism of displacement in some anxiety disorders and undoing in OCD.

Cognitive-behavioural theories of anxiety point to the role of conditioning and socialization processes in the development of anxiety disorders. Family systems theories highlight the roles of family belief systems and interaction patterns in the development and maintenance of anxiety disorder and the significance of family lifecycle transitions in precipitating the onset of these conditions. In clinical practice many clinicians take an integrative approach to assessment and treatment of anxiety disorders.

4

Depression

Feelings of happiness and sadness are adaptive. Many behaviour patterns that lead to happiness, such as socializing with others and developing longstanding friendships, are important for the survival of the species. Sadness, which commonly follows loss of valued objects, events, personal characteristics and skills or relationships, may also be adaptive, in that it reminds us to take care of valued objects, events, characteristics or relationships in future, lest we lose them again. Extreme mood states such as mania and depression are less adaptive. There is no doubt that, during periods of mania or hypomania, some individuals with bipolar disorder—which is characterized by episodes of mania and depression—produce highly creative work (Jamison, 1995). However, this is done at a high cost. Inevitably people who suffer from bipolar disorder, during a manic phase, run the risk of exhaustion, dehydration and excessive and dangerous risk-taking. Seasonal affective disorder (or winter depression as it is colloquially known) may be linked phylogenetically to hibernation and this may have been adaptive for our cave-dwelling ancestors.

However, nowadays depression seems to fulfil no adaptive function. Despite this, it is a highly prevalent condition. Because of its prevalence, the main focus in this chapter will be on major depression, although reference will be made to other mood disorders such as bipolar disorder. Depression is not simply 'feeling sad'. Major depression is a recurrent episodic condition involving low mood; selective attention to negative features of the environment; a pessimistic belief-system; self-defeating behaviour patterns, particularly within intimate relationships; and a disturbance of sleep and appetite.

After considering the clinical features and epidemiology of depression, theoretical explanations for mood problems will be presented later in this chapter. Each of these specific explanations has been developed within the context of one of four broad theories. These are the biological, psychodynamic, cognitive-behavioural and family systems theories of psychological problems. Later these four broad theories are reviewed with reference to their main attributes, their contributions to our understanding and treatment of psychological problems, and their limitations.

Case Example

May, a single woman in her early thirties, was referred for counselling by her GP. She insisted on being seen by the psychologist at her house, since she had not been out of bed for two years. May had her first episode of depression in her mid-twenties after her first and only significant adult relationship with a man (Rob) ended. The episode lasted almost a year and consequently she lost her job as a teacher. During this first episode of depression she spent the time in bed. She lived in a small maritime town about 250 miles from the village where she grew up. Her mother came to stay with her during her first episode of depression. At her family doctor's suggestion she also attended a series of counselling sessions which resulted in her recovery. Shortly after this the counsellor left the district. After a few months and a series of disappointing and unsuccessful attempts to rekindle old friendships, May relapsed. She spent two years in bed and refused to see anyone except her mother and the family doctor.

Family History. May was born and brought up until the age of 8 in a small rural village. She moved house when she was 8 and lost a number of close friends through the change in address. This was particularly stressful because she had difficulty making new friends. She also relied on her old friends a lot to make living with her parents bearable. So when she moved house, she lost this support. Her home life was very stressful because she was regularly beaten brutally by her father and knew that her mother was also physically abused by him. This continued throughout her life until she left home to go to college at 18. During her childhood and teenage years she spent a lot of time at home studying, to distract her from the unhappy home atmosphere.

On the positive side, she had a good relationship with her cousins and some happy childhood memories about staying at their house. She had a couple of good friends at college with whom she went on holidays to Greece. When she left college, she began work as a teacher and loved her job. She got on well with children and was admired for this by her colleagues. Within her extended family there was a history of mood and alcohol problems. Her aunt and a cousin had both suffered from depression. She also had an uncle with a drink problem.

Presentation. May presented with feelings of sadness, emptiness and an inability to experience pleasure. She talked about herself in self-deprecating ways. For example, she reported the following beliefs: 'I'm no good as a teacher. I've lost my job because of this illness; I'm no good as a woman. I'll never be married. I'm no good as a person, I'm dirty and worthless and I'm rotting inside.' She also viewed the world negatively. What follows are some of her beliefs: 'My father is no good. He beat me as a child and beat my mother. He is the reason why I am ill. I can't change the past so I will be ill forever. My mother is no good. I would recover if she were not here looking after me. She interferes in my life and tries to control me. I have no friends so there is no point in recovery. Whatever pleasant things that I have experienced were few and far between. For example, my holiday in Greece. I had no control over either the good or bad things that happened to me, so I cannot control my recovery.' May's view of the future was also bleak. For example, she said: There is no point in recovery because other people will only take advantage of me. I have been unfortunate in the past so I will always be unfortunate.'

Notable features of May's behaviour were the fact that she lived a constricted housebound lifestyle; experienced early morning waking, diurnal variation of mood, and appetite and weight loss; and refused medication.

Formulation. May's family history of mood disorders suggests that she may have had a genetic vulnerability which, at a biological level, predisposed her to developing depression. Her negative childhood experiences, particularly physical abuse, moving house, losing supportive friends and using homework as a way to distract herself from her stressful family life, probably led her to develop a set of core beliefs, assumptions and coping styles that rendered her vulnerable to depression at a psychological level. These assumptions included the following: 'I deserve to be hurt. I am only acceptable to myself and others if I am hurt. Other people deserve to be hurt. You can't trust anyone because they will abandon you. If people don't like me, I'm worthless. If I fail at work, I'm worthless.' Each of her episodes of depression was triggered by a precipitating stressful life event. The first episode was triggered by her losing her boyfriend and the second by her failing to rekindle old friendships, the loss of her job and possibly the loss of her counsellor.

These stressful events reactivated all of her depressive beliefs and assumptions so she was prone to view the world in negative terms. This negative way of interpreting events maintained her depressed mood. She also had a cognitive style which maintained her depressed mood. She tended to minimize positive events and maximize negative events, and attribute failure experiences to personal characteristics rather than situational factors. In addition to these cognitive maintaining factors, aspects of her behaviour also maintained her depressed mood, particularly her constricted life-style. Her lifestyle reduced her opportunities for forming relationships, finding work or experiencing pleasure. The absence of these opportunities confirmed her negative view of herself, her world and her future. She was locked into a vicious cycle. Eventually this had taken its toll on her appetite, circadian rhythms and activity level. She slept poorly, awoke early, ate little and rarely exercised. These sleeping, eating and activity problems may also have maintained her depressed mood. These difficulties may also have been subserved by dysregulation of neurotransmitters in the midbrain.

Intervention. Following assessment, May was helped to under-stand the formulation. She engaged in a multimodal treatment programme involving cognitive-behavioural interventions, family therapy and antidepressant medication. Cognitive therapy helped her to challenge her depressive thinking style and view the world in more positive terms. Behaviour therapy helped her alter her self-defeating patterns of behaviour and expand her constricted lifestyle. Family systems therapy helped her parents, particularly her mother, reduce their inappropriate overinvolvement with May and apologize for abuse to which May had been subjected as a child. Anti-depressant medication targeted the neurophysiological dysregulation which was presumed to underpin her depression. Over a period of months May increased her activity level, developed a more positive thinking style, and greater autonomy from her parents, and began to engage in a more normal lifestyle.

Clinical Features

The main features of depression are presented in Table below. These features may be linked by assuming that depressed individuals have usually suffered a loss of some sort—either a loss of an important relationship, a loss of some valued attribute such as athletic ability or health, or a loss of status.

With respect to perception, having suffered a loss, depressed individuals tend to perceive the world as if further losses were probable. Depressed people selectively attend to negative features of the environment and this in turn leads them to engage in depressive cognitions and unrewarding behaviour patterns which further entrench their depressed mood. In severe cases of depression, individuals may report mood-congruent auditory hallucinations. We may assume that this severe perceptual abnormality is present when individuals report hearing voices criticizing them or telling them depressive things. Auditory hallucinations also occur in schizophrenia. However, the hallucinations that occur in schizophrenia are not necessarily mood-congruent.

With respect to cognition, depressed individuals describe themselves, the world and the future in negative terms. They evaluate themselves as worthless and are critical of their occupations and social accomplishments. Often this negative self-evaluation is expressed as guilt for not living up to certain standards or letting others down. They see their world, including family, friends and work or school, as unrewarding, critical and hostile or apathetic. They describe the future in bleak terms and report little if any hope that things will improve. Where they report extreme hopelessness and this is coupled with excessive guilt for which they believe they should be punished, suicidal ideas or intentions may be reported.

Table.

Clinical features of depression

Perception	• Perceptual bias towards negative events • Mood-congruent hallucinations*
Cognition	• Negative view of self, world and future

(Cont..)

	• Excessive guilt • Suicidal ideation* • Mood-congruent delusions* • Cognitive distortions • Inability to concentrate
Affect	• Depressed mood • Inability to experience pleasure • Irritable mood • Anxiety and apprehension
Behaviour	• Psychomotor retardation or agitation • Depressive stupor*
Somatic state	• Fatigue • Disturbance of sleep • Aches and pains • Loss of appetite or overeating • Change in weight* • Diurnal variation of mood (worse in morning) • Loss of interest in sex
Interpersonal adjustment	• Deterioration in family relationships • Withdrawal from peer relationships • Poor school performance

Note: Features are based on ICD 10 and DSM IV descriptions of major depression. * These features occurs in severe episodes of depression

Extremely negative thoughts about the self, the world and the future may be woven together in severe cases into depressive delusional systems. In addition to the content of the depressed individual's thoughts being bleak, they also display logical errors in their thinking and concentration problems. Errors is reasoning are marked by a tendency to maximize the significance and implications of negative events and minimize the significance of positive events. Concentration and attention difficulties lead to difficulties managing occupational, academic or leisure activities demanding sustained attention.

With respect to affect, low mood is a core feature of depression. Depressed mood is usually reported as a feeling of sadness, loneliness or despair and an inability to experience pleasure. Alternatively, irritability, anxiety and aggression may be the main features, with sadness and inability to experience pleasure being less prominent. This is not surprising since normal grief is characterized by sadness at the absence of the lost object, anger at the lost object for abandoning the grieving person and anxiety that further losses may occur.

At a behavioural level, depressed individuals may show either reduced and slowed activity levels (psychomotor retardation) or increased but ineffective activity (psychomotor agitation). They may show a failure to engage in activities that would bring them a sense of achievement or connectedness to family or friends. Where individuals become immobile, this is referred to as depressive stupor. Fortunately this is rare.

Somatic or vegetative features such as loss of energy, disturbances of sleep and appetite, weight loss or failure to make age-appropriate weight gain, abdominal pains or headaches, and diurnal variation in mood are all associated with more severe conditions. A loss of interest in sex may also occur. These features of depression are consistent with findings that dysregulation of neurophysiological, endocrine and immune functions are associated with depression and that sleep architecture is also affected. This material will be mentioned in more detail in the section on biological theories of depression.

At an interpersonal level, depressed individuals report a deterioration in their relationships with family, friends, work colleagues and other significant figures in their lives. They describe themselves as lonely and yet unable or unworthy to take steps to make contact with others. One complication of depression is self-destructive behaviour. In classifying self-destructive behaviour, a distinction is made between suicidal and parasuicidal behaviour. With suicidal behaviour, the intention to kill oneself is the aim of the self-destructive act. With parasuicide, the person may hope to resolve an interpersonal difficulty by making a self-harming gesture. For example, the person may hope to elicit care, concern or pity. Risk factors and triggering events associated with suicide are set out.

Classification

Within DSM IV and ICD 10 distinctions are made between:

- Major depression
- Bipolar mood disorder
- Dysthymia
- Cyclothymia

Major depression and bipolar disorder are both episodic mood disorders, with the former being characterized by episodes of low mood, negative cognition, sleep and appetite disturbance, and the latter being characterized in addition by episodes of mania in which elation, grandiosity, flight of ideas and expansive behaviour occur. Dysthymia and cyclothymia are non-episodic chronic conditions, with dysthymia being characterized by depressive symptomatology and cyclothymia being characterized by similar but less extreme mood fluctuations than bipolar disorder. The distinctions between unipolar and bipolar conditions and between recurrent and persistent disorders currently used in ICD 10 and DSM IV have replaced distinctions used in earlier classification systems. These distinctions include:

- Neurotic and psychotic depression
- Endogenous and reactive depression
- Overt and masked depression

Reviews of the classification of mood disorders identify the following reasons for abandoning these earlier distinctions (Kendell, 1976; Farmer and McGuffin, 1989; Harrington, 1993). The neurotic and psychotic distinction, based originally on inferred psychodynamic etiological factors and differences in observable symptoms, has been discarded because evidence for inferred psychodynamic etiological differences has not been supported by empirical evidence. The endogenous-reactive distinction has been abandoned because evidence from stressful life event research shows that almost all episodes of depression, regardless of their quality or severity, are preceded by stressful life events and in that sense are reactive. The recognition that youngsters with depression may show comorbid conduct disorders has rendered the concept of masked depression unnecessary, since the term was often used in child and adolescent psychology to classify depressed youngsters who *masked* their low mood with angry outbursts of aggressive or destructive behaviour.

Epidemiology

The lifetime prevalence of major depression is 10-25 per cent for women and 5-12 per cent for men. For bipolar disorder the lifetime prevalence for men and women is about 1 per cent (APA, 1994). Depression is less common among children and adolescents than it is among adults (Harrington, 1993). In community samples prevalence rates of depression in preadolescence range from 0.5 to 2.5 per cent and in adolescents from 2 to 8 per cent. Depression is equally common in preadolescent boys and girls but more common in post-adolescent females than males.

Comorbid dysthymia, anxiety disorders, substance abuse disorders, eating disorders and borderline personality disorder are common in major depression (APA, 1994). Early onset, comorbid dysthymia (often called double depression), severe depressive symptoms, maternal depression, and the absence of comorbid conduct problems during episodes of childhood depression have all been shown in longitudinal studies to be predictive of worse outcome (Harrington, 1993). Up to 15 per cent of people with major depression and bipolar disorder commit suicide.

Etiological Theories

Theoretical explanations for depression and related treatments have been developed within biological, psychoanalytic, cognitive-behavioural, and family systems traditions. A number of influential theories from each of these areas will be briefly reviewed below.

BIOLOGICAL THEORIES

Biological theories of depression point to the role of genetic factors in rendering people vulnerable to the development of mood disorders, and to dysregulations of neurotransmitter, endocrine and immune systems or dysregulations of biorhythms as central to their etiology.

Genetic theories. Results of twin studies and family studies suggest that a predisposition to mood disorders may be genetically transmitted (Andrew *et al.*, 1998). Precisely what biological characteristics are genetically transmitted and the mechanisms of transmissions are unknown. However, the results of studies conducted on amine dysregulation, endocrine abnormalities, immune system dysfunction, circadian rhythm abnormalities and the seasonal occurrence of depression in some cases suggest that a biological vulnerability to dysregulation of one or more of these systems is probably inherited. It is also probable that the vulnerability is polygenetically transmitted since the results of family studies cannot easily be accounted for by simpler models of genetic transmission. The evidence that bipolar disorder is genetically transmitted is particularly compelling.

Amine dysregulation theories. Amine dysregulation theories argue that hypoactivity of the amine systems in neuroanatomical centres associated with reward and punishment is central to depression (Deakin, 1986). Available evidence shows that depression occurs when there is a dysregulation of the amine systems in the medial forebrain bundle (reward system) and the periventricular system (punishment system) which affect drives for seeking pleasure and appetite. Noradrenalin and serotonin are the main neurotransmitters involved. Originally depletion of these neuro-transmitters was thought to cause depression, but now a more complex dysregulation of the system involving a reduction in the sensitivity of postsynaptic receptor sites is hypothesized to be the critical difficulty.

Amine dysregulation theories predict that psychopharmaco-logical treatment which increases the sensitivity of the dysfunctional postsynaptic membranes or increases the amount of serotonin or noradrenalin in the reward and punishment centres of the brain will lead to recovery. Three main classes of antide-pressant medication have been used to test these predictions. Tricyclic antidepressants (TCAs) like imipramine/Tofranil increase the sensitivity of dysfunctional receptor sites to neurotransmitters, particularly noradrenalin. Monoamine oxidase inhibitors (MAOIs) like phenelzine/Nardil prevent the enzyme—monoamine oxidase—from breaking down neurotransmitters in the synaptic cleft and lead to an increase in amine levels. Selective serotonin reuptake inhibitors (SSRIs) like fluoxetine/Prozac prevent serotonin from being reabsorbed into the presynaptic membrane and so increase levels of this neurotransmitter. Results of treatment trials show that all three classes of antidepressants are effective in alleviating depression in up to two-thirds of adult cases (Nemeroff and Schatzberg, 1998) but are remarkably ineffective in alleviating depression in children or adolescents (Gadow, 1992).

Endocrine dysregulation theories. Abnormalities of those aspects of the endocrine system associated with thyroxin and cortisol are posited as the cause of depression in endocrine dysregulation theories (Deakin, 1986). One line of evidence shows that in depressed adults a dysregulation of the hypothalamic-pituitary-thyroid axis which governs thyroid levels occurs which leads thyroxin levels to fall below normal. Possibly, low levels of thyroxin are responsible for some aspects of depression. A second line of evidence points to a dysregulation of the hypothalamic-pituitary-adrenal axis which governs release of cortisol under stress. The occurrence of depression in some instances follows exposure to chronic (but not acute) stress which in turn is associated with elevated cortisol levels and abnormal cortisol circadian rhythms. In comparison with non-depressed individuals, people with depression show a more rapid recovery over a twenty-four-hour period in response to dexamethasone (a synthetic cortisollike compound). Research continues on using the dexamethasone suppression test (DST) as a biological aid to diagnosing major depression, although results of studies with children and adolescents have not been consistent with those obtained for adult populations (Harrington, 1993).

Immune system dysfunction theories. A variety of theorists have argued that there is a link between the functioning of the immune system and the occurrence of depression (Levy and Heiden, 1991). A growing body of evidence shows that exposure to chronic stress or acute loss such as bereavement leads to both impaired immune system functioning and depressive symptoms. Impaired immune system functioning increases susceptibility to infections and consequent illnesses, which may be perceived as additional life stresses. These in turn may maintain or exacerbate depression.

Circadian rhythm desynchrony theories. These theories argue that depression occurs when there is a dysregulation or desynchrony in the circadian rhythms which govern the sleep-waking cycle. In support of these theories, sleep studies of depressed people show that they have abnormal circadian rhythms characterized by shortened rapid eye movement (REM) onset latency, broken sleep, early morning waking and difficulties with sleep onset (Kupfer and Reynolds, 1992). The neuroanatomical basis for this desynchrony may lie in the reticular activating system which has been shown to govern arousal and the sleep-waking cycle. This line of research has shown that in some instances depression may be diagnosed by assessing REM onset latency, which is reduced in those with depression. Furthermore some depressed people temporarily recover following sleep deprivation and this has led to the exploration of REM deprivation as a potential treatment for depression.

Seasonal rhythm dysregulation theories. It has also been found that some people show depressive symptoms in wintertime (Wehr and Rosenthal, 1989). This condition is known as seasonal affective disorder. One hypothesis about seasonal affective disorder is that this is a phylogenetic derivative of hibernation. Symptoms of fatigue, oversleeping and increased appetite are under the control of melatonin which is secreted by the pineal gland during periods of diminished daylight. Artificially lengthening the day using light therapy is the most important treatment to derive from this line of research. There is also a line of research at present investigating the effects of administering melatonin at key times to alter the time of its release from the pineal gland and thereby ameliorate depressive symptoms.

Psychoanalytic Theories

Of the many psychoanalytic theories of depression that have been developed, reference will be made here to Freud's (1917) original position, Bibring's (1965) ego-psychological model and Blatt's object relations formulation (Blatt and Zuroff, 1992). These theories have been selected because they are illustrative of psycho-dynamic explanations, and Blatt's model has been singled out for attention because, unlike many psychodynamic theories, considerable effort has gone into empirically testing it.

Freud's classical psychoanalytic theory. In Freud's (1917) psychoanalytic theory it was argued that following protracted separation or bereavement (referred to in psychoanalysis as object loss) regression to the oral stage occurs, during which a distinction between self and the lost object is not made and the lost object is introjected. Subsequent aggression at the introject of the lost object for bringing about a state of abandonment is experienced as self-directed anger or the self-hatred which characterizes depressed people. People whose primary caregivers either failed to meet their dependency needs during the oral phase or who were overindulgent are predisposed to developing depression as described within this model. This is because they devote much of their energy to desperately seeking love by working hard or devoting themselves to helping others, at great personal cost. When they lose a loved one, they feel the loss more acutely and are more likely to regress, introject the lost object and experience retroflexive anger.

Freud also made provision for losses such as unemployment which could symbolize object losses within his theory. Because the superego, which is not fully developed in children, is the psychological structure necessary for directing anger at the ego, the traditional psychoanalytic position entails the view that children are unable to experience depression. This view is unsupported by available epidemiological data, reviewed above. However, Freud's position was important in drawing attention to the significance of self-directed anger in the maintenance of depression in some cases. Freud also pointed out the importance of early life experiences in creating a vulnerability to depression, an idea which is central to modern psychodynamic and cognitive theories of depression.

Bibring's ego-psychology theory. Bibring (1965), a later psychodynamic ego psychologist, explained depression as the outcome of low self-esteem which resulted from perceiving a large discrepancy between the self as it is and the ideal self. Internalization of harsh critical parental injunctions or perfectionistic parental injunctions during early childhood accounted for the development of a particularly unrealistic ego-ideal. A substantial body of evidence supports the view that low self-esteem is an important correlate and in some instances precursor of depression, and in some but not all cases this is associated with a history of critical or punitive parenting (Blatt and Zuroff, 1992).

Blatt's object relations theory. Blatt and Zuroff (1992) argue that there are two types of depression associated with two distinct types of early parent-child relationships which engender vulnerability to depression when faced with two distinct types of stresses in later life. A vulnerability to stresses involving *loss of attachment* relationships is central to one type of depression and this has its roots in early experiences of neglectful or overindulgent parenting. A vulnerability to stresses involving *loss of autonomy* and control is central to the other type of depression and this has its roots in early experiences of critical, punitive parenting. This distinction between depression associated with disruption of interpersonal relationships and that associated with threats to mastering important achievement-oriented tasks has been made by many psycho-dynamic object relations theorists, and indeed by cognitive theorists including Beck, whose theory will be discussed below, but has found its clearest articulation in Blatt and Zuroff's work. A growing body of evidence shows that, in adults, these subtypes of depression are associated with the recall of different childhood experiences, which presumably have led to the development of different types of depressive object relations which are activated in later life by different types of stressful life events.

According to Blatt and Zuroff, children who receive either neglectful or overindulgent parenting develop internal working models for later life relationships in which expectations of abandonment are a central feature. For such individuals, denial and repression are the most common defence mechanisms employed to deal with perceived threats. These individuals are particularly vulnerable in later life to stressful events that involve the disruption of relationships, such as rejection or bereavement. When they develop a mood disorder it is characterized by a preoccupation with the themes of abandonment, helplessness and a desire to find someone who will provide love. On the other hand, children exposed to critical and punitive parenting develop internal working models for relationships in which the constructs of success and failure or blame and responsibility are central organizing features. Projection or reaction formation are the most common defences used by such individuals. In teenage years and adulthood such individuals are particularly vulnerable to experiences of criticism, failure or loss of control. Their mood disorders are characterized by a sense of self-criticism, inferiority, worthlessness and guilt.

Within the psychodynamic tradition, individual psychodynamic psychotherapy or classical psychoanalysis are the treatments of choice (Bateman and Holmes, 1995). In such therapies, transference develops so that self-directed hostility, self-criticism, and ideas of abandonment or loss of autonomy are projected onto the therapist. This transference is interpreted. The therapist points out parallels between the patient-parent relationship and the patient-analyst relationship. Also, parallels between these relationships and those that patients have with other significant people in their lives are noted. This process of interpretation helps patients learn to identify when they are falling into problematic relationship habits in future. The analytic relationship also provides patients with a forum where they can ventilate and work through the intense depressive and angry feelings that underpin their problematic ways of managing relationships. This frees them to explore more realistic standards for self-evaluation and to develop more trusting internal working models for relationships. There is some evidence from controlled trials that brief psychodynamic psycho-therapy is an effective treatment for people with depressive symptomatology (Holmes, 1999).

Cognitive and Behavioural Theories

Of the many behavioural theories of depression, those of Lewinsohn (Lewinsohn and Gotlib, 1995) and Rehm (Kaslow and Rehm, 1991) are particularly important because they have led to a considerable amount of research on the effectiveness of behavioural treatments for depression. Beck (1976) and Seligman's (Abramson, Seligman and Teasdale, 1978) cognitive theories of depression are among the most important and influential in the field. They have spawned an extraordinary amount of research on psychological processes in depression and the effectiveness of cognitive therapy (Williams *et al.*, 1992). For these reasons, the theories of Lewinsohn, Rehm, Beck and Seligman will be considered in this section.

Lewinsohn's behavioural theory. In Lewinsohn's behavioural theory, he argues that depression is maintained by a lack of response-contingent positive reinforcement (RCPR) (Lewinsohn and Gotlib, 1995). This may occur because people with depression lack the social skills required for eliciting rewarding inter-actions from others. Treatment programmes based on this model include individual or group social-skills training which aims to train clients in the skills necessary for receiving response-contingent positive reinforcement and to arrange the environment so that there are many opportunities for using these social skills. Controlled evaluations of such programmes with adults and analogue adolescent clients support their effectiveness.

Rehm's self-control theory. While Lewinsohn's theory focuses on the roles of environmental contingencies in depression, Rehm bases her model of depression on a consideration of internal contingencies. According to Rehm's self-control theory, depression arises from deficits in self-monitoring, self-evaluation and self-reinforcement (Kaslow and Rehm, 1991). Specifically, depression arises when a person selectively monitors the occurrence of negative events to the exclusion of positive events; selectively monitors immediate rather than long-term consequences of actions; sets overly stringent criteria for evaluating actions; makes negative attributions for personal actions; engages in little self-reinforcement for adaptive behaviours; and engages in excessive self-punishment. Treatment programmes derived from this model aim to improve the skills required for more effective self-monitoring, self-evaluation and self-reinforcement. Controlled evaluations of such programmes with adults and analogue adolescent clients support their effectiveness.

Beck's cognitive theory. According to Beck's (1976) theory, depression occurs when life events involving loss occur and reactivate negative cognitive schemas formed early in childhood as a result of early loss experiences. These negative schemas entail negative assumptions such as *I am only worthwhile if everybody likes me*. When activated, such schemas underpin the occurrence of negative automatic thoughts, such as *no one here likes me*, and cognitive distortions, such as *all or nothing thinking*. Negative schemas have their roots in loss experiences in early childhood including:

- Loss of parents or family members through death, illness or separation
- Loss of positive parental care through parental rejection, criticism, severe punishment, overprotection, neglect or abuse
- Loss of personal health
- Loss or lack of positive peer relationships through bullying or exclusion from peer group
- Expectation of loss, for example, where a parent was expected to die of chronic illness

According to Beck, two negative schemas, which contain latent attitudes about the self, the world and the future, are of particular importance in depression. The first relates to interpersonal relationships and the second to personal achievement. He referred to these as sociotropy and autonomy. Individuals who have negative self-schemas where sociotropy is the central organizing theme define themselves negatively if they perceive themselves to be failing in maintaining positive relationships. Thus their core assumption about the self may be *If I am not liked by everybody, then I am worthless.* Individuals who have negative self-schemas where autonomy is the central organizing theme define themselves negatively if they perceive themselves to be failing in achieving work-related goals. Thus their core assumption about the self may be *If I am not a success and in control, then I am worthless.* When faced with life stresses, individuals vulnerable to depression because of early loss experience and the related development of negative self-schemas become prone to interpreting ambiguous situations in negative mood-depressing ways. The various logical errors that they make are referred to by Beck as cognitive distortions and these include the following:

- *All or nothing thinking.* Thinking in extreme categorical terms—for example, either I'm a success or a failure.
- *Selective abstraction.* Selectively focusing on a small aspect of a situation and drawing conclusions from this—for example, I made a mistake so every thing I did was wrong.
- *Overgeneralization.* Generalizing from one instance to all possible instances—for example, he didn't say hello so he must hate me.
- *Magnification.* Exaggerating the significance of an event—for example, he said she didn't like me so that must mean she hates me.
- *Personalization.* Attributing negative feelings of others to the self—for example, he looked really angry when he walked into the room, so I must have done something wrong.
- *Emotional reasoning.* Taking feelings as facts—for example, I feel like the future is black so the future is hopeless.

Depressed individuals interpret situations in terms of their negative cognitive schemas and so their automatic thoughts are characterized by these depressive cognitive distortions. Automatic thoughts are self-statements which occur without apparent volition when an individual attempts to interpret a situation so as to respond to it in a coherent way.

Cognitive therapy aims to train clients to monitor situations where depressive assumptions and distortions occur; to evaluate the validity of these depressive assumptions and distortions; and to engage in activities that provide evidence to refute the negative assumptions. With adults, cognitive therapy is as effective as antidepressant medication in alleviating depressive symptoms and more effective in relapse prevention than medication therapy (Craighead *et al.*, 1998a). It has also been shown to be effective with adolescents (Harrington *et al.*, 1998).

Seligman's reformulated learned helplessness theory. According to Seligman, depression arises when a person repeatedly fails to control the occurrence of aversive stimuli or failure experiences and makes internal, global, stable attributions for these failures and external, specific, unstable attributions for success (Abramson *et al.*, 1978). Attributional retraining in which clients learn to attribute failure to external, specific, unstable factors and success to internal, global, stable factors is the principal treatment to emerge from this model. In addition the individual's environment may be modified so that the likelihood of non-aversive successful experiences greatly outweighs the likelihood of aversive failure experiences. Clients may also be trained to reduce their preferences for success experiences that are beyond their abilities (Seligman, 1981). The success of Beck's cognitive therapy and that of Lewinsohn's behavioural therapy, mentioned above, both lend support to this model.

Family Systems Theories

Family systems theories of depression highlight the importance of family-based stress, support, belief systems and interaction patterns in the etiology and maintenance of depression. What follows is one example of such a theory which we have elaborated in greater detail elsewhere (Carr, 2000b). Both genetic and environmental factors contribute to the development of depressive conditions. The amount of stress required to precipitate the onset of an episode of depression is proportional to the genetic vulnerability. That is, little stress may precipitate the onset of an episode in individuals who are genetically vulnerable to the condition, whereas a great deal of stress may be necessary to precipitate the disorder in individuals who have no family history of depression.

A variety of family factors may predispose people to developing depression in later life. Both loss experiences and exposure to stresses in early life, which prevent needs for safety and security being met, may render individuals vulnerable to depression in later life. Loss experiences include unsupported separations from primary caregivers; parental psychological absence, for example, through depression; and loss of a parent or primary caregiver through bereavement. Stresses which prevent needs for safety and security being met include child abuse, neglect, multi-placement experiences, parental conflict, family disorganization and long-term exposure to a pessimistic family culture.

Episodes of major depression may be precipitated by stressful family lifecycle transitions. Loss experiences associated with the disruption of significant relationships and loss experiences associated with failure to achieve valued goals, in particular, may precipitate an episode of depression. For example, marital relationships may be disrupted through conflict and criticism; infidelity and violations of trust; physical and psychological abuse; or threats of separation. Failure to achieve valued goals and threats to autonomy may occur with exam failure, work-related performance difficulties or unemployment.

Depression may be maintained by particular types of personal and family belief systems, notably those characterized by a preoccupation with past losses, a negative view of the self as valueless and powerless, and a pessimism and hopelessness about the future. Such depressive belief systems may lead to a reduction in activities and an avoidance of participation in relationships that might disprove these depressive beliefs or lead to a sense of pleasure and optimism.

Depression may be maintained by particular patterns of family interaction, and nowhere is this more evident than in marital relationships where one partner is depressed. The depressed partner behaves more and more helplessly and in response the other partner engages in more and more caretaking, so that the entire relationship becomes defined in terms of these two rigid complementary positions. Depressed partners have difficulty fulfilling their routine duties at home and work, and so some of these may be taken on by the non-symptomatic partner. Depressed partners typically provide and elicit little support or sexual fulfilment within their marriages, and in this sense non-symptomatic partners suffer a major loss of support when their partners develop depression. Depressed partners are less able to engage in effective joint problem-solving and this is frustrating for their partners who may find that important joint decisions are left unmade or are made unsatisfactorily. Depressed partners continually seek both reassurance and confirmation of their negative views of themselves, a set of conflicting demands that is aversive for their partners and may lead to distancing.

The development of this complementary behaviour pattern greatly compromises the couple's capacity to meet each other's needs for desired levels of intimacy and autonomy. For both partners, the need for personal power and autonomy is not met. The depressed person believes that he or she is helpless to change the situation because of intrinsic powerlessness or because the world is too bleak or dangerous. Caregiving partners experience themselves as trapped in an endless and futile round of caregiving where nothing they do make their partner better and yet they feel compelled to continue caregiving. This frustration of their need for autonomy gives rise to anger which neither partner may believe is appropriate to express. The symptomatic partner may believe that it would be ungrateful to criticize their partner for excessive ineffectual caregiving. Caregiving partners may believe that it would be insensitive to criticize their symptomatic partner for not recovering.

However, periodically either partner may become so frustrated that they express their intense anger at their partner. In these instances, depressed individuals find such aggression from a previously supportive partner exacerbates their depression. Subsequently, guilt for expressing aggression may lead them to return to their previous roles of apparently grateful care-receiver or apparently dutiful caregiver. This type of behaviour pattern prevents couples from meeting each other's needs for psychological intimacy. They are only able to view each other as caregivers or receivers and unable to accept each other as people who are quite distinct from the problem and who are jointly facing the challenge of managing the depression.

Over time, this type of caregiving and receiving behaviour pattern may deteriorate into one where more frequent verbal criticism, aggression or distancing and infidelity occur. In other cases these hostile responses to depression are there from the start. Verbal and physical aggression, distancing, infidelity and threatened separation all confirm the depressed partner's belief system concerning the hopelessness and dangerousness of the world and so maintain the depression. The exacerbated symptoms may elicit further aggression or distancing from the non-symptomatic partner. However, extremely depressive and helpless behaviour has been found to inhibit non-symptomatic partners' expression of verbal or physical aggression. So, in some couples, the depressed spouse learns that one way to avoid being attacked verbally or physically is to show extreme symptoms. This display of extreme symptoms also has a payoff for the non-symptomatic partner in so far as it inhibits aggression and so prevents the occurrence of the guilt which follows aggressive displays.

This family systems conceptualization of depression in adults is supported by evidence on the role of stress and social support in the genesis of depression; the importance of negative interaction patterns characterized by criticism in its maintenance; and the effectiveness of therapies that alter problem-maintaining patterns of family interaction (Joiner and Coyne, 1999; Clarkin *et al.*, 1988; Craighead *et al.*, 1998a; Prince and Jacobson, 1995; Baucom *et al.*, 1998). Interpersonal therapy which helps individuals renegotiate conflictual role relationships has been shown to be as effective as cognitive therapy for depression. Behavioural marital therapy has been shown to effectively alleviate depression, particularly in cases where there is comorbid marital distress.

An Integrative Approach

In clinical practice clinicians often adopt an integrative approach to understanding and treating depression. People may be predisposed to develop depression due to a genetic vulnerability and/or early loss or stressful life experiences. Depression may be precipitated by stress or lifecycle transitions. It may be maintained by biological, psychodynamic, cognitive, behavioural or family inter-actional factors. Probably the most effective treatment programmes are multimodal and involve antidepressant medication to address the neurobiological difficulties that underpin the syndrome, coupled with psychological therapy to address psychodynamic, cognitive, behavioural or family interactional factors (Nemeroff and Schatzberg, 1998; Craighead *et al.*, 1998a; Prince and Jacobson, 1995; Baucom *et al.*, 1998).

For bipolar mood disorders, long-term treatment with either lithium carbonate or anticonvulsants (such as carbamezapine) has been shown to lead to clinically significant mood stabilization (Keck and McElroy, 1998). However, stabilization is not complete and relapses occur. These are associated with alterations in family and work-related stress and support and adherence to medication regimes (Craighead *et al.*, 1998b). Clarkin *et al.* (1988) found that psychoeducational family therapy improved clients' long-term adjustment. The psychoeducational intervention provided family members with information on bipolar disorder as a chronic illness; helped them develop ways to reduce life stress and increase support for the patient; and encouraged them to maximize medication adherence.

CONTROVERSIES

There are many controversies in the scientific study and clinical treatment of mood disorders. The finding that twice as many women as men are diagnosed with depression, whereas in child-hood the prevalence of depression is the same for boys and girls, has led to a variety of conflicting viewpoints (Brems, 1995). One view is that the higher rate of depression in women arises from the societal inequalities that favour men over women and the oppressive sex-roles into which women fall when they enter long-term heterosexual relationships. A second view is that, with the transition to adolescence, females undergo biological changes that render them vulnerable to depression. A third view is that women are more likely than men to acknowledge and report personal distress and so receive a diagnosis of depression. Men, in contrast, are more likely to deny personal distress and deal with it by using alcohol, recreational drugs, or immersion in work. Because of this they are less likely to be diagnosed with depression. However, they have higher rates of alcoholism, drug abuse and heart disease. There is some evidence to support all of these positions. The issue of gender and depression remains a major controversy in the study of mood disorders.

The medicalization of mood problems and personal distress is a second controversy deserving particular mention

(Breggin, 1991; Kutchins and Kirk, 1999). The arguments here are similar to those outlined earlier in relation to anxiety and fear. Within ICD 10 and DSM IV, mood disorders are framed as medical conditions requiring treatment, and in clinical practice pharmacological treatment is commonly favoured because it is cheaper and easier to offer than non-pharmacological alternatives. In cases where depression is unresponsive to medication it is common practice to offer patients a course of electroconvulsive therapy (ECT). An alternative viewpoint is that this way of conceptualizing personal distress and grief further disempowers people who are already feeling powerless. For example, if people find that their mood is persistently low, their thoughts are self-critical and pessimistic, they are unable to regulate their sleeping and eating routines, they are unable to initiate rewarding or valued activities without extreme effort, and they find their important relationships are deteriorating, it may lead them to believe that they are truly powerless to control their lives if these experiences are defined as an illness requiring pharmacological treatment or ECT. Furthermore, antidepressant medication and ECT have side-effects that may be very distressing for some people. With antidepressants, dry mouth, blurred vision and stomach upsets may occur (Nemeroff and Schatzberg, 1998). With ECT, memory loss and confusion typically occur (Lock, 1999).

Those who are critical of the medicalization of mood problems would argue that if a person can understand that personal distress arising from loss of valued objects, events, personal skills, characteristics and achievements, and important relationships underpins the changes in mood, thought, behaviour and relationships that are labelled as depression, they may be empowered to take action to make changes in their routine ways of thinking, acting and managing their relationships. This type of action may involve developing their social support networks, engaging in regular exercise, joining self-help groups or participating in counselling or psychotherapy. At a broader level, those who criticize the 'pill for every ill' philosophy that underpins the medicalization of mood problems would argue that the high prevalence of depression is a symptom of structural difficulties within modern industrialized society. The real cure—for what we call depression—may involve reducing the ridiculously high standards that we measure our valued characteristics, skills and achievements against; creating safer environments to reduce losses due to crime, injury and accidents; developing community structures, family structures and work practices that foster the development of social support networks; and promoting an optimistic world view which values people taking personal responsibility for both their own well-being and the well-being of their communities (Seligman and Csikszentmihalya, 2000).

Evidence cited in this chapter shows that some people are genetically vulnerable to mood disorders and mood disorders commonly entail a unique constellation of biological, psychological and social features which are absent in non-depressed people. Furthermore, the most effective treatments for depression are multimodal and involve both biological and psychological components. In so far as the use of diagnostic categories as working hypotheses supports the development of a more refined under-standing of extreme mood states and their management, diagnoses such as major depression or bipolar disorder are valuable. However, it is also valuable to study the psychosocial contexts in modern industrialized society within which personal distress evolves and those psychosocial contexts which foster resilience in the face of loss or adversity. It is valuable also to study the social processes that underpin the medicalization and the medical treatment of mood problems in clinical practice and to explore the degree to which this and alternative psychosocial conceptualizations of mood problems empower clients to take control of their lives and avoid extreme mood states.

Major depression is a recurrent episodic condition involving low mood; selective attention to negative features of the environment; a pessimistic belief-system; self-defeating behaviour patterns, particularly within intimate relationships; and a disturbance of sleep and appetite. Loss is often the core theme linking these clinical features: loss of an important relationship, loss of some valued attribute such as health, or loss of status, for example, through unemployment. Major depression is distinguished from bipolar disorder where there are also episodes of elation, and from dysthymia which is a milder non-episodic mood disorder. Depression is a relatively common disorder and up to 15 per cent of people with major depression commit suicide. Theoretical explanations for depression have been developed within biological, psychoanalytic, cognitive-behavioural, and family systems traditions. In clinical practice an integrative approach to understanding depression may be adopted and a multimodal approach to treatment in which biological and psychological treatments are combined.

5

Schizophrenia

Schizophrenia- A group of disorders characterized by loss of contact with reality, marked disturbances of thought and perception, and bizarre behavior. At some phase delusions or hallucinations almost always occur. Schizophrenia is among the most debilitating and complex of the psychoses. Approximately 1% of the world population is afflicted with this mental illness. "Emil Kraepelin first identified the illness in 1896 when he distinguished it from the mood disorders. Kraepelin believed that all psychiatric disorders were caused by organic factors, and his experience suggested to him that the onset of the disease occured early in the life of the individual. Hence, he called it *dementia praecox*, which means a premature deterioration of the brain."

Emil's thoughts were later disputed by many psychiatrists. One of these was Eugene Bleuler, an eminent Swiss psychiatrist, who, in 1911 found that the onset of the disease could in fact occur in the later years. He also reported that schizophrenia was not characterized by the progressive deterioration over the life of the patient, but rather that most patients, after an original severe deterioration, tend to stabilize and remain at the same point in their psychosis for extended periods of time. Bleuler also felt that in order to avoid any misunderstanding of the nature of the illness by the now obvious misnomer attached to it, the disease would be much better served if it was referred to as "schizophrenia."

Bleuler invented the word by combining two Greek words meaning "split" and "mind." This emphasized a splitting apart of the patient's affective and cognitive functioning, which are heavily affected by the disease.

Types

There are two types of schizophrenia accordingly enumerated— Type I (Reactive or Acute Schizophrenia) and Type II (Process Schizophrenia):

Reactive or Acute Schizophrenia

Reactive schizophrenia is usually sudden and seems to be a reaction to some life crisis. Since the premorbid history is usually good, when the disease does manifest itself, it is in the early phases. Reactive schizophrenia is a more treatable form of the illness than process or chronic schizophrenia.

Process Schizophrenia

Also reffered to as poor premorbid schizophrenia, this type is characterized by lengthy periods of its development with a gradual deterioration and exclusively negative symptoms. It doesn't seem to be related to any major life change or negative event. Usually this type of schizophrenia is associated with "loners" who are rejected by society, tend not to develop social skills and don't excel out of high school.

Symptoms

Positive Versus Negative

The positive symptoms are things like bizarre behavior, hallucinations, or delusions. Negative symptoms refer to the absence of any adjustive behavior in the important areas of life, a chronic maladaptiveness, flatness of affect, and absence of developed interpersonal relationships (social skills).

Paranoid Versus Nonparanoid

These symptoms are more easily defined as the presence of heavily paranoid behavior or the absence thereof.

The presence of paranoid symptoms early in the disease apparently suggests a good prognosis. There is a relationship between reactive schizophrenia and paranoid thinking.

Content of Thought

The principal disturbance in the schizophrenic's thought processes is multiple delusions. This is divided into two sub-categories, persecutory delusions (in which the schizophrenic believes that he/she is being talked about, spied upon, or their death being planned) and delusions of reference (which is when the schizophrenic gives personal importance to completely unrelated incidents, objects, or people. Other common delusions include thought broadcasting (they believe their thoughts are visible to the outside world) and thought insertion, which is what most people perceive schizophrenia as consisting of (their thoughts are not their own and are in truth being inserted into their minds by some outside force). Other delusions, such as believing oneself to be Jesus, may appear in extreme cases.

Form of Thcught

Either schizophrenics express their thoughts in a loose manner, where ideas shift from one subject to another with seemingly no purpose, or "poverty of content," where communication is so vague, abstract, or repetitive, that it is meaningless to the listener. Made up words or illogically stringed together phrases may appear in writing or speech as well.

Perception

As we well know, the perception of the world is distorted in the experience of a schizophrenic. This may occur with any of the afflicted's senses, but most often appear as auditory, with voices in the patients head or commands from high authorities which are obeyed at high risk to others or the patient themselves. Visual hallucinations happen less often.

Affect

This symptom is easiest described as an excessive lack of corellation between what an individual is saying and what emotion they are expressing (e.g. recounting an experience of serious horror while chuckling).

Volition

Simply the occurence of paralyzation on the patient's will to act out on a decision by their ambivalence.

Sense of Self

Schizophrenics generally are not aware of their individuality to an extent that they maintain a perplexity about who they are.

Relationship to the External World

Although obvious, most schizophrenics are so preoccupied with the effects of their illness that they tend to be unavailable to others, which is reffered to as autism. They don't notice the world that is happening before them.

Classes of Schizophrenia

Paranoid Schizophrenia: Patient displays the psychotic symptoms.

Undifferentiated Schizophrenia: Used when the patient's symptoms clearly point to schizophrenia but are so clouded that classification into the different types of schizophrenia is very difficult.

Residual Schizophrenia: Advised when an individual has been through at least one episode of schizophrenia (6 months) but then "recover."

Schizophreniform Disorder - Best understood as a schizophrenic disorder that has lasted for more than two weeks but less than six months. A less serious diagnosis, as it has a likelihood for the patient to return as normal member of society.

The term Schizophrenia refers to a collection of seriously debilitating conditions characterized chiefly by hallucinations, delusions and thought disorder. Hallucinations involve experiencing a sensation in the absence of an external stimulus. For example, with auditory hallucinations people report hearing voices that others cannot hear. Delusions are unfounded and culturally alien beliefs. For example, with persecutory delusions individuals may believe that a group of people are conspiring to harm them. Thought disorder refers to disorganized and illogical speech. When negative symptoms are present, the person may live a restricted lifestyle involving very little activity, and little social interaction with others and may express a very narrow range of emotions. People diagnosed with schizophrenia may also show negative symptoms such as restriction of activity, speech and affect.

After considering the clinical features and epidemiology of schizophrenia, theoretical explanations of this condition will be presented later in the chapter. Each of these specific explanations has been developed within the context of one of three broad theories. These are the biological, cognitive-behavioural and family systems theories of psychological problems.

Case Example

Julian was referred for assessment and advice by his GP. His parents were worried about him because he had been behaving strangely since returning to the family home after studying in London for a year. Julian had failed his exams and came home, he said, 'to sort his head out'. He lacked concentration and his conversation was incoherent much of the time. Also his behaviour was erratic. He had gone jogging one morning the previous week and not returned. His parents had found him in his training shoes and running shorts 35 miles away later that day. He was exhausted and dehydrated.

He explained that he was on his way to the car ferry to Holland on a secret mission.

Developmental and Family History. Julian was the 19-year-old son of a prominent local farmer. His father was devoted to his work and had a positive, if distant, relationship with Julian. His mother was a painter and also a prominent local figure. She was a warm flamboyant woman who retained strong anti-establishment beliefs. She responded to Julian in the interview setting with intense concern and protectiveness. There was no family history of psycho-logical disorder but some members of the mother's family were well-to-do but odd or eccentric, especially her brother Sedrick and her uncle William Junior. William's eccentricities led him into serious conflict with his father and Sedrick's odd behaviour underpinned his highly conflictual childless marriage to Patsy.

Julian grew up on the farm and went to the local school. His development was essentially normal. He was excellent at cricket and did well at school. He had many friends locally and in college where he had been for a year. He had engaged in some experimental drug abuse having taken some LSD and smoked cannabis fairly regularly over a six-month period. He had found the previous three months at college difficult and had an intense fear of exam failure.

Presentation. Julian presented with delusions, hallucinations and disorganized speech. He was reluctant to be interviewed because he believed he had urgent business to attend to in Holland. He explained that his path was to the east where he was being called by an unknown source. He knew this because of the sign he had seen that morning when out jogging. The way the wheel of the old cart caught the sunlight and made a pattern on the wall, he interpreted as a clear sign for him to go. When he questioned this idea, a clear authoritative voice said that he should leave at once. When asked to continue his story, he had blocked and lost the thread of what he was saying. He began to giggle. He said that he could hear someone say something funny. Later he said that he must go soon because people would try to prevent him. He had heard them talking about him the day before. They had tried to put bad ideas into his head. He described being frightened by this and by periodic sensations that everything was too loud and too bright and coming at him. He said, 'it was like doing acid [LSD] all the time'.

Formulation. In this case the principal predisposing factors were a possible genetic vulnerability (because of eccentric maternal family history) and hallucinogenic drug use. Among the important precipitating factors were his transition to college and exam pressure. His condition was maintained by a high level of maternal expressed emotion (principally overinvolvement).

Treatment. The treatment plan included neuroleptic medication and family work to reduce parental expressed emotion.

Clinical Features

The clinical features of schizophrenia are presented in Table below. Psychotic episodes may last from one to six months, although some extend to a year. They are usually preceded by a prodromal period of a number of weeks. Psychotic episodes may be shortened and the severity of symptomatology ameliorated through early detection and the use of pharmacological and psychological treatment as outlined below (McGorry, 1998). Inter-episode functioning may vary greatly and better inter-episode functioning is associated with a better prognosis. The duration of remission between episodes may be lengthened through the use of maintenance medication, family intervention to reduce the amount of stress to which the individual is exposed, and the use of stress-reducing coping strategies.

At a perceptual level individuals with schizophrenia describe a breakdown in perceptual selectivity, with difficulties focusing on essential information or stimuli to the exclusion of accidental details or background noise. In a florid psychotic state, internal stimuli (or thoughts) are interpreted as originating from another source and experienced as auditory hallucinations.

Table.

Clinical features of schizophrenia

Perception	• Breakdown in perceptual selectivity • Hallucinations
Thought	• Formal thought disorder • Delusions • Impaired judgement and reality testing • Confused sense of self
Emotion	• Prodromal anxiety and depression • Inappropriate affect • Flattened affect and impoverished affect • Postpsychotic depression
Behaviour	Prodromal sleep disturbance • Prodromal impulsivity • Prodromal repetitive compulsive behaviour • Impaired goal-directed behaviour • Catatonia, negativism and mutism
Interpersonal adjustment	• Poor school or work performance • Withdrawal from peer relationships • Deterioration in family relationships

Note: Features are based on ICD 10 and DSM IV descriptions of schizophrenia.

Individuals may perceive voices to vary along a number of dimensions. Voices may be construed as benign or malevolent; controlling or impotent; all-knowing or knowing little about the person; and the person may feel compelled to do what the voice says or not. Hallucinations that are perceived to be malevolent, controlling, all-knowing and which the individual feels compelled to obey are far more distressing than those that are not construed as having these attributes.

At a cognitive level, formal thought disorder occurs in schizophrenia and is characterized by a difficulty following a logical train of thought from A to B. Judgement may be impaired and unusual significance may be accorded to unrelated events so that delusions occur. Delusions, from a cognitive-behavioural perspective, are beliefs or inferences that have been drawn to explain why a particular set of events has occurred. Delusions may vary in the degree of conviction with which they are held (from great certainty to little certainty); the degree to which the person is preoccupied with them (the amount of time spent thinking about the belief); and the amount of distress they cause. Particular sets of delusions may entail a confused sense of self, particularly paranoid delusions where individuals believe that they are being persecuted or punished for misdeeds, or delusions of control where there is a belief that one's actions are controlled by others.

At an emotional level, during the prodromal phase, anxiety or depression may occur in response to initial changes in perceptual selectivity and cognitive inefficiency, and a key part of relapse prevention is learning how to identify and manage prodromal changes in affect. During the florid phase, high arousal levels may occur in response to the experience of delusions and hallucinations. Inappropriate affect may be present, particularly in hebephrenic schizophrenia, where the individual responds not to the external social context but to internal stimuli such as auditory hallucinations. Flattened affect may also occur, particularly in chronic cases where high levels of medication have been taken for extended time periods. In response to an episode of psychosis, the sense of loss may give rise to postpsychotic depression.

At a behavioural level, during the prodromal phase, sleep disturbance, impulsive behaviour, and compulsive behaviour may be present. During psychotic episodes, goal-directed behaviour is impaired and, in chronic cases, negativism, mutism and catatonia may occur. At an interpersonal level a deterioration in relationships with family members may occur. Social withdrawal from interaction with peers may occur and at work or school there is usually a marked decline in performance.

Epidemiology and Classification

The prevalence of schizophrenia is about 1 per cent in populations over 18 years of age. This is an international cross-cultural finding replicated in Ireland by NíNualláin *et al.* in the 1980s, although, prior to NíNualláin's (1984) work, methodologically-flawed studies suggested that the incidence of schizophrenia in Ireland was higher than in other countries. The marked variability among people with schizophrenia in symptomatology, course, treatment response and possible etiological factors has led to the development of a variety of classification systems. Also, many psychotic conditions which closely resemble schizophrenia have been identified. In ICD 10 and DSM IV, symptomatology, rather than inferred biological or psychological etiological factors, is used as a basis for subtyping schizophrenia. Four main subtypes may be distinguished:

- Paranoid
- Catatonic
- Hebephrenic or disorganized
- Undifferentiated

Where paranoid delusions predominate, a diagnosis of paranoid schizophrenia is given. Cases in which psychomotor abnormalities predominate are classified as catatonic. People with catatonic schizophrenia may alternate between extremes of excitability and stupor; they may show automatic obedience or negativism; they may adopt peculiar postures for long periods; and they may show waxy flexibility of the limbs. Cases are classified as hebephrenic in the ICD 10 and disorganized in the DSM IV when inappropriate or flat affect is the principal feature and where there is disorganization of behaviour and speech. In both ICD 10 and DSM IV, when cases do not fall into the three categories just mentioned, they are classified as undifferentiated. In clinical practice, it is often difficult to define the boundaries between schizophrenic disorders and other conditions, particularly the following:

- Mood disorders
- Personality disorders (particularly schizoid and schizotypal)
- Drug-induced psychoses
- Pervasive developmental disorders such as autism

The boundaries between schizophrenic disorders and mood disorders have been dealt with in the ICD and DSM systems by proposing mixed disorders such as schizoaffective disorder or postpsychotic depression. Comorbid depression is common in schizophrenia and about 10 per cent of people with schizophrenia commit suicide (Kopelowicz and Liberman, 1998). Where aspects of symptomatology are insufficient for a diagnosis of schizophrenia, ICD 10 and DSM IV indicate that diagnoses of personality disorders may be given. Less extreme levels of social withdrawal and aloofness than those typical of people with schizophrenia are the primary characteristic of the schizoid personality disorder. People showing eccentricities which fall just short of thought disorder and delusions may be diagnosed as having schizotypal personality disorder in DSM IV (although this condition is classified along with psychotic disorders in ICD 10). Extended periods of observation and a referral for toxicological tests will usually throw light on cases of drug-induced psychoses. Autism may be distinguished from schizophrenia by the absence of sustained delusions and hallucinations. The following characteristics may be predictive of a good outcome where people are diagnosed with schizophrenia (Asarnow, 1994; Mari and Streiner, 1994; Neale and Oltmanns, 1980):

- Good premorbid adjustment
- Rapid onset
- Precipitating stressful life events

- Family history of affective disorder (rather than schizophrenia)
- Additional affective features in the case's presentation
- Favourable life situation to return to
- Low incidence of family psychopathology
- Adolescent (rather than child)
- Female (rather than male)

Etiological Theories

Historically research on schizophrenia has followed from two principal traditions: the first represented by Kraepelin (1896) and the second by Bleuler (1911). Kraepelin defined the condition as principally characterized by a large constellation of observable symptoms (such as delusions, hallucinations and thought disorder) and a chronic course, due to an underlying degenerative neurological condition. In contrast, Bleuler defined schizophrenia in terms of a disturbance in a circumscribed set of inferred psychological processes. He speculated that the capacity to associate one thought with another, to associate thoughts with emotions, and the self with reality, was impaired or *split*—hence the term schizophrenia (from the Greek words for *split* and *mind*). He argued that observable symptoms such as delusions and hallucinations were secondary to these central psychological difficulties.

For Bleuler, the symptoms of schizophrenia such as delusions and hallucinations represented the person's attempt to cope with the world despite disruption (or *splitting*) in central psychological processes. Up until the late 1970s, Bleuler's tradition, associated with a broad definition of schizophrenia, predominated in the US, whereas in the UK, Ireland and Europe, Kraepelin's narrower definition held sway. Following the landmark US-UK diagnostic study (US-UK Team, 1974) that highlighted the extraordinary differences between the way schizophrenia was defined in

America and the way it was defined in Britain, there has been a gradual move towards developing an internationally acceptable set of diagnostic criteria. The narrowing of the gap between the North American and European definitions of schizophrenia is reflected in the marked similarity between the diagnostic criteria for the disorder contained in ICD 10 and DSM IV. The two definitions include delusions, hallucinations, disorganized speech and bizarre behaviour as central to the definition of schizophrenia. They also include negative symptoms which involve a restriction of activity, speech and affect as important features of the condition.

Modern research on schizophrenia has also been guided by two broad groups of theories. The first, in the tradition of Kraepelin, has been concerned largely with the role of biological factors in the etiology and maintenance of the disorder. The second group of theories, in the tradition of Bleuler, has addressed the role of psychological factors in schizophrenia. These psychological theories fall into two broad categories. First, there are those which highlight intrapsychic difficulties and deficits, which are of critical importance for cognitive behavioural interventions in schizophrenia. Second, there are those which point to the important role of social factors in the etiology of schizophrenia. These have informed the development of family systems interventions.

In addition to these unifactoral theories, an increasing number of integrative diathesis-stress models of schizophrenia have been developed which attempt to offer integrative explanations which take account of a myriad of biological and psychological factors in the development and course of this very puzzling condition. Brief summaries of representative hypotheses from each of these groups of explanations will be presented below, along with comments on the degree to which the various positions are supported by available evidence.

Biological Theories

Genetic, neurophysiological and neuroanatomical factors have all been implicated in the etiology of schizophrenia. These factors have been studied within the context of the theories outlined in this section.

Genetic hypothesis. The genetic hypothesis argues that a biological predisposition which is genetically inherited renders some individuals vulnerable to schizophrenia. The results of the most carefully controlled genetic studies to date indicate that the lifetime risk for developing schizophrenia is proportional to the amount of shared genes. For monozygotic twins the risk is 48 per cent; for the children of two parents with schizophrenia the risk is 46 per cent; for dizygotic twins or children with one affected parent, the risk is 17 per cent; for grandchildren the risk is 5 per cent; and for members of the general population the risk is 1 per cent (Gottesman, 1991). The genetic mechanisms of transmission are unknown and a focus for current gene mapping and chromosomal research. The majority of people who develop schizophrenia have no relatives who suffer from the condition and many people who have a possible genetic predisposition to schizophrenia do not develop the syndrome. It is therefore likely that some intrauterine, social or physical environmental factors must also contribute to the development of schizophrenia.

The dopamine hypothesis. The dopamine hypothesis suggests that schizophrenia is caused by a dysfunction of the mesolymbic dopaminergic system (Snyder, 1986). There is controversy over the precise nature of this dysfunction. An excess of dopamine, an excess of dopamine receptors, or a supersensitivity of postsynaptic dopamine receptors are the principal dysfunctional mechanisms for which there is some empirical support. The theory is supported by studies of the pharmacological treatment of schizophrenia which show that the neuroleptic medication, such as chlorpromazine or clozapine, blocks dopamine activity in people with schizophrenia (Davis *et al.*, 1991).

Treatments with first-generation antipsychotic drugs such as chlorpromazine, haloparadol, flupenthixol or second-generation antipsychotic preparations such as clozapine, risperidone, olanzapine and sertindome are the main approaches to pharmacological intervention for psychotic conditions deriving from the dopamine hypothesis (Sheitman *et al.*, 1998). While these antipsychotic agents control positive symptoms of schizophrenia in the majority of cases, they are not without problems. They have short-term side-effects such as parkinsonism which is often controlled by an anti-parkinsonism agent such as cogentin. Tardive dyskinesia, an irreversible neurological condition, is one of the tragic long-term side-effects of neuroleptic drug usage. For this reason, ideally, the lowest possible dose of neuroleptic medication should be used (Perry *et al.*, 1997).

From a theoretical perspective, it is unlikely that the dopamine theory is sophisticated enough to account for the symptoms of schizophrenia and its response to antipsychotic medication. Contrary to predictions derived from the dopamine hypothesis, antipsychotic agents such as chlorpromazine do not act immediately in reducing symptoms, although they immediately affect dopamine levels. They only affect positive symptoms (such as delusions and hallucinations) and have little impact on negative symptoms (such as inactivity or poverty of speech). They rarely completely eliminate positive symptoms. Nor are they effective for all people with schizophrenia. About 25 per cent of people with schizophrenia do not respond to antipsychotic medication. In the neurophysiological domain, the dopamine hypothesis will ultimately be replaced by a far more complex hypothesis involving a number of neurotransmitter systems including serotonin. However, as yet proposals are still in their infancy.

Neurodevelopmental hypothesis. The neurodevelopmental hypothesis argues that factors in the prenatal intrauterine environment (such as a maternal flu virus) and perinatal neurological insults (often indexed by a history of obstetric complications) lead to atypical neuroanatomical development which culminates in the emergence of schizophrenia (Murray and Lewis, 1987). A growing body of evidence offers some support for this position. Three neuroanatomical abnormalities have consistently emerged in neuroimaging and postmortem studies of schizophrenia and in particular in studies of cases with predominantly negative symptoms (McGlashen and Fenton, 1992). The first abnormality concerns the reduced size of the frontal lobes and reduced levels of neurological functioning in the frontal lobes. This abnormality may underpin the peculiarities of gait, posture, patterns of eye-movements and related attention and memory deficits that have been consistently found to typify the condition in laboratory studies. The second abnormality is the enlargement of the cerebral ventricles (particularly the left ventricle) associated with atrophy of the brain. The third abnormality is neuronal degeneration in the cortex. These neuroanatomical abnormalities may account for the soft neurological signs that are particularly common among patients showing negative symptoms. Accumulating evidence suggests that maternal viral infections and obstetric complications may contribute to the development of neuro-anatomical abnormalities that occur in some forms of schizophrenia (Werry and Taylor, 1994).

Two syndrome hypothesis. The two syndrome hypothesis argues that a distinction may be made between type 1 schizophrenia, which is a genetically inherited disease, marked by a dysregulation of the mesolymbic dopaminergic system and characterized by positive symptoms, and type 2 schizophrenia, which is a neurodevelopmental disorder arising from pre- or perinatal insults and marked by chronic negative symptoms (Crow, 1985). Type 1 schizophrenia, it is suggested, has an acute onset, clear precipitants, a good response to antipsychotic medication and good inter-episode adjustment. Poor premorbid functioning, an insidious onset, a chronic course, neuropsychological deficits and a poor response to medication are thought to characterize type 2 schizophrenia. The two syndrome hypothesis is an attempt to draw together and integrate the lines of research that have followed from the genetic hypothesis, the dopamine hypothesis and its derivatives, on the one hand, and the neurodevelopmental hypothesis, on the other. The two syndrome hypothesis fits a good proportion of available data but is probably an oversimplification, since many cases show aspects of both syndromes (Werry and Taylor, 1994; Lieberman and Koreen, 1993; McGlashen and Fenton, 1992).

COGNITIVE THEORIES

Cognitive theories have focused on the cognitive deficits and biases present in schizophrenia.

Cognitive deficit hypothesis. Cognitive deficit theory argues that schizophrenia is characterized by a core cognitive deficit which underpins the many symptoms that define the disorder, including delusions, hallucinations, thought disorder, inactivity, poverty of speech and so forth. Cognitive theories fall within the tradition which began with the work of Bleuler (1911) who attempted to explain the symptoms of schizophrenia in terms of the breaking of associative threads. More recent cognitive deficit hypotheses have attempted to explain schizophrenia in terms of attentional deficits and these types of explanations have been supported by laboratory studies which confirm that people with schizophrenia perform poorly on the following information processing and perceptual tasks: reaction time, visual tracking, short-term recall, competing information tasks, size estimation, visual masking and categorization (Neale and Oltmanns, 1980).

A particularly sophisticated cognitive deficit hypothesis has been offered by Hemsley (1996) who proposes that the core difficulty in schizophrenia is a breakdown in the relationship between stored information and current sensory input. Usually, stored memories or regularities, such as particular social situations like entering a shop to buy groceries, influence which aspects of a situation are attended to and which are ignored. Such stored information also influences our expectancies and interpretation of the situation. This process normally occurs automatically and rapidly. According to Hemsley, for people with a diagnosis of schizophrenia, this rapid automatic orientation to a new context does not occur. The similarity of the new situation to other similar situations is not automatically recognized. There is no clarity about which internal or external stimuli to attend to and which to ignore.

The theory offers a useful framework for understanding the clinical features of schizophrenia. Auditory hallucinations, within this frame of reference, are a result of the sensory overload that arises from this filtering problem. Internal events are misinterpreted as sensations arising from external stimulation. Inappropriate affect, such as giggling accompanied by unusual gesturing, may occur in response to such hallucinations. Delusional belief systems, according to Hemsley's theory, reflect efforts to impose meaningful relationships on the barrage of highly confusing external and internal stimuli and hallucinations that enter consciousness without filtering. For example, tactile hallucinations may be interpreted as receiving messages from the TV. Early on in schizophrenia, such attempts may be sufficient to lead to the development of an internally consistent belief system, as in the case of paranoid schizophrenia. However, with time, it may become too difficult to retain internal consistency in the belief system and so there may be a later progression to non-paranoid schizophrenia. Impaired judgement and reality testing due to sensory overload may prevent adolescents with schizophrenia from checking the validity of delusional beliefs against the facts.

Psychophysiological studies show that, in the prodromal period before an acute psychotic episode, patients become hyper-aroused by ambient environmental stress and also become hyper-reactive to certain social and environmental stimuli (Hemsley, 1996). This hyperarousal disrupts the perceptual and cognitive processes which are already to some degree impaired. If habituation does not occur, and arousal increases beyond a critical threshold, then disruption of perceptual and cognitive processes continues and a florid psychotic episode will occur. The experience of sensory overload entailed by the inability to select out the essential from the incidental may lead to a breakdown in the capacity for logical thinking. An inability to coherently give an account that follows logical progression occurs and this is referred to as formal thought disorder.

Negative symptoms such as impoverished or flattened affect, impoverished speech, impaired goal-directed behaviour and social withdrawal may all reflect attempts to cope with the intense experience of sensory overload and a complete breakdown in the attentional filter that is normally employed to orient us to different types of situation and screen out irrelevant internal and external stimuli. Antipsychotic medication and the psychological therapies described below aim to restore the capacity to manage sensory input, reduce the complexity of this input, and re-establish routines of living.

Cognitive bias theory. The cognitive-behavioural approach to schizophrenia, rather than addressing the syndrome as a whole, offers discrete explanations for individual symptoms or discrete methods for conducting therapy to alter specific symptoms (Haddock and Slade, 1996). A tenet held by many cognitive-behavioural theorists is that psychotic symptoms do not form a syndrome which reflects an underlying disease or disorder. They also deny that alterations in psychological functioning which occur in people with a diagnosis of schizophrenia are discontinuous with normal psychological processes. Rather, psychotic experiences are viewed as being on a continuum with normal experiences. Thus delusions are not meaningless, but are strongly-held irrational beliefs, the formation of which has been influenced by cognitive biases, in the same way, for example, as depressive beliefs are formed in mood disorders.

Chadwick, Birchwood and Trower (1996) have offered a clinically useful theoretical analysis of hallucinations and delusions in terms of an ABC framework. Activating events trigger Beliefs based on inferences and attributions and these ways of thinking about the activating events lead to emotional and behavioural Consequence. They argue that hallucinations (which are Activating events) are loud thoughts that have been misattributed to other sources and about which inferences of power and need for compliance have been made, and these Beliefs lead to distressing emotions and behaviour which are a Consequence of the beliefs. The cognitive therapy of hallucinations involves challenging beliefs about the power of the voices and the importance of complying with the commands of the voices. When these are altered through collaborative empirical testing in therapy, then the emotional distress and related behaviour abate. Delusions are explained as Beliefs or inferences which have been constructed to make sense of particular Activating events and lead to various negative Consequences including distressing emotions and behaviour. When these are altered through collaborative empirical testing in therapy, then the emotional distress and related behaviour abate. A growing body of evidence shows that treatment based upon this type of analysis of psychotic symptoms ameliorates targeted symptoms (Haddock and Slade, 1996).

Prodromal hypothesis. The prodromal hypothesis argues that individuals with schizophrenia experience a prodromal set of symptoms, largely perceptual hypersensitivity and cognitive information processing deficits, which herald the onset of a psychotic episode. These prodromal symptoms are exacerbated by inferences and attributions made by the individual about the controllability and origins of their unusual prodromal perceptual and cognitive experiences, and these attributions increase arousal and accelerate the onset of the relapse. Cognitive behavioural interventions made during the prodromal period which target attributions, or pharmacological interventions which target alterations in perceptual and information processing functions, should, according to the theory, prevent relapse or reduce the severity of relapse.

Clinical studies have identified four stages in the development of a psychotic episode (Birchwood, 1996). In the first stage there is a feeling of loss of control over cognitive and perceptual processes as a breakdown in perceptual selectivity occurs. This may be accompanied by a feeling of heightened awareness and mental efficiency and yet an inability to prevent internal and external events from invading consciousness. A sense of anxiety (a fear of going crazy) may occur at this point. In the second stage, depression characterized by low mood, low self-esteem, social withdrawal, poor school performance and vegetative features such as sleep disruption occurs in reaction to the deterioration of cognitive processes. Some individuals try to cope with this deterioration by engaging in compulsive rituals that will give them a sense that they can impose order on what is an increasingly chaotic experience.

In the third stage, disinhibition occurs and individuals act impulsively, giving free rein to aggression, self-destruction, sexual urges, wishes to travel and so forth. This impulsivity may lead adolescents, particularly, to create social situations in which they become exposed to high levels of stimulation which in turn may precipitate the onset of florid psychotic symptoms. For example, becoming involved in fights, atypical sexual encounters, or impulsively travelling a long distance may all lead to complex unpredictable and highly stressful experiences, which will be perceived as all the more stressful because of the breakdown in perceptual selectivity.

In the fourth stage, prepsychotic thinking occurs, with frequent perceptual misinterpretations and delusional explanations given for them. Often these delusional explanations involve ideas of reference or paranoid ideation. Thus, as perceptual processes become more dysfunctional, individuals continue to try to make sense of their very unusual experiences by developing beliefs that are at variance with the culture and their usual belief systems.

It appears that each patient has their own unique 'relapse signature', with specific experiences occurring in a unique order, but following this broad four-stage model. Learning the pattern of this prodromal phase can improve relapse management. Studies show that the shorter the duration of the untreated illness, the less likely the patient is to relapse in the subsequent two-year period (Birchwood, 1996).

Social and Family Systems Theories

Social and family systems theories of schizophrenia, which have implications for family systems treatment approaches, focus on the role of proximal and distal social stresses, especially their effects on the course of the disorder.

Social class and social drift hypotheses. The social class hypothesis argues that stresses associated with low socio-economic status probably contribute to the genesis of schizophrenia. The social drift hypothesis, in contrast, argues that people diagnosed as having schizophrenia are probably ill-equipped to deal with the pressures associated with upward mobility and so drift into a lower socio-economic bracket. This in turn may create stresses which exacerbate their coping problems. Available evidence supports both hypotheses. There is an inverse relationship between social class and schizophrenia and, following diagnosis, social drift occurs (Dohrenwend *et al.*, 1992). However, the relationship between social class and schizophrenia is not a specific relationship. Many other psychological problems are more common in lower socio-economic groups.

Stressful life events theory. That schizophrenia is a reaction to a build-up of life stress is the core tenet of this theory. Stressful life events may precipitate a schizophrenic episode. Studies reviewed by Neale and Oltmanns (1980) show that in the three-week period before an episode up to two-thirds of cases with schizophrenia reported stressful life events whereas only a fifth of matched controls with other diagnoses experienced at least one stressful life event.

Family environment theory. Recent theories about the role of the family in schizophrenia argue that regular intense, confusing or threatening interactions with family members may precipitate a relapse, increase the need for antipsychotic medication or exacerbate symptomatology during a psychotic episode (Perlmutter, 1996). The implication is that the impaired perceptual selectivity and cognitive information processing deficits associated with schizophrenia make intense, confusing or threatening social stimuli very difficult to cope with. A further implication of this type of theory is that interventions that help family members to reduce the amount of intense, confusing or threatening interactions to which the youngster is exposed will reduce the relapse rate and need for medication.

Expressed emotion, a combined measure of criticism and overinvolvement obtained in an individual interview with a parent, has been shown to be related to the course of schizophrenia in adults and adolescents (Mari and Streiner, 1994). While most current work on the family environment and schizophrenia focuses on the expressed emotion construct, the constructs of communication deviance and affective style have both been shown to be more prevalent in families with an adolescent member with a diagnosis of schizophrenia (Doane *et al.*, 1981). Affective style is rated from transcripts of family interaction, and more families containing a member with a diagnosis of schizophrenia show predominantly critical or intrusive interaction patterns. Communication deviance is assessed on the basis of parental responses to ambiguous inkblot stimuli from the Rorschach test administered in a one-to-one context, but is interpreted as being reflective of family members' difficulties in maintaining a focus when engaged in joint problem-solving. With communication deviance an amorphous pattern of thinking may occur where there is little differentiation and communications are vague, or a fragmented pattern where there is much differentiation but little integration of ideas. There is a growing consensus that expressed emotion and affective style probably reflect parental responses to the disorganized and bizarre behaviour characteristic of schizophrenia. There is also some consensus that communication deviance may reflect a genetic trait which finds expression as a subclinical thought disorder process in the parents.

The strongest evidence for the family environment hypothesis comes from intervention studies which have shown that reducing the level of criticism and overinvolvement in families with schizophrenic members reduces the need for medication; greatly prolongs the inter-episode interval; and significantly reduces the relapse rate (Tarrier, 1996). The results of six major studies show that family interventions reduce relapse rates at two years from about 70 per cent to about 25 per cent in families with high expressed emotion containing a member with a diagnosis of schizophrenia (Mari and Streiner, 1994).

An Integrative Approach

While early research on schizophrenia was guided by simple single-factor theories, since the publication of Zubin and Spring's paper in 1977, 'Vulnerability: A new view of schizophrenia'—there have been numerous attempts to provide overarching multifactorial models of schizophrenia which provide a framework for integrating the results of research from a variety of biological and psychosocial perspectives (Zubin and Ludwig, 1983; Nuechterlin and Dawson, 1984). Typically these have been termed diathesis-stress models. They argue that, for the symptoms of schizophrenia to occur, a biologically vulnerable individual must be exposed to environmental stress. The interaction of the vulnerability factors with the stress factors leads to the occurrence of the symptomatology. This is subsequently maintained by ongoing exposure to environmental stress and by the way in which the person reacts to this stress and copes with the unusual experiences associated with schizophrenia. For clinicians and researchers alike, the challenge is to work out the vulnerabilities and stresses present in each case and the processes of interaction linking the two classes of factors. A diathesis-stress model of schizophrenia is presented in Figure below. The model shows how genetic factors and intrauterine adversities may lead to a biological vulnerability to schizophrenia. This vulnerability may take the form of neurophysiological and neuro-anatomical abnormalities with the latter being predominantly associated with negative symptoms and the former associated with positive symptoms.

This biological vulnerability underpins a psychological vulnerability which includes perceptual and cognitive information processing deficits. As part of the process of attempting to manage these unusual perceptual and cognitive experiences, cognitive distortions and misattributions may occur which lead to a perception of internal stimuli (such as hallucinations) and external stimuli as threatening, as evidenced by anxiety-provoking delusions. The cognitive deficits and distortions lead to hyper-arousal. With exposure to stressful life events, stresses associated with low socio-economic status, and stress associated with the family's response to prepsychotic or psychotic symptomatology, hyperarousal and cognitive deficits and distortions may become exacerbated. This may lead in the first instance to the onset of a psychotic episode or, subsequently, to the maintenance of psychotic experiences or relapse. This type of model has clear implications for assessment and intervention.

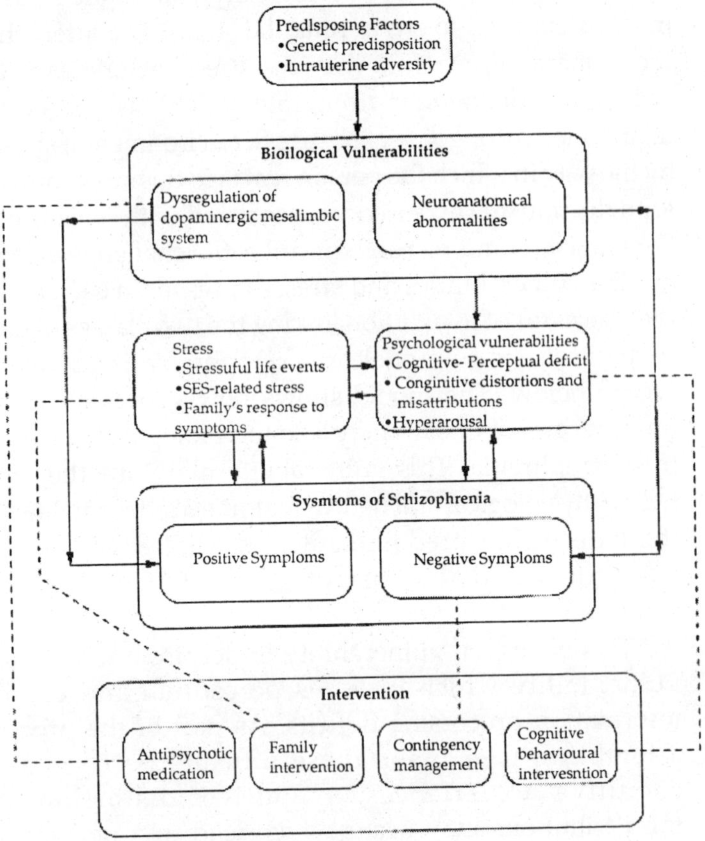

Figure. *A diathesis-stress model of schizophrenia*

It suggests that multi-modal programmes are the treatment of choice and this viewpoint is supported by available empirical evidence (Sheitman *et al.*, 1998;

Kopelowicz and Liberman, 1998). Multimodal programmes should combine antipsychotic medication, family systems interventions, cognitive behavioural therapy and behavioural interventions. Antipsychotic medication may target the dysregulation of the dopaminergic system. Family interventions may be used to reduce the stressfulness of family interactions. Cognitive behavioural interventions may be employed to facilitate coping with positive symptoms. Contingency management may be used to increase activity levels and alter negative symptoms.

Schizophrenia and Split Personality

In popular culture schizophrenia is often inaccurately used to refer to split personality. However, in psychology, it is clear from this chapter that schizophrenia does not refer to such a condition. The closest scientific equivalent to split personality is referred to as multiple personality disorder (MPD) in ICD 10 and dissociative identity disorder (DID) in DSM IV. The central feature is the apparent existence of two or more distinct personalities within the individual, with only one being evident at a time. Each personality (or alter) is distinct, with its own memories, behaviour patterns and interpersonal style. Commonly the host personality is unaware of the existence of the alters and these vary in their knowledge about each other.

A diathesis-stress model of dissociative identity disorder is now widely accepted (Maldonado *et al.*, 1998). Available evidence suggests that the capacity to dissociate is normally distributed within the population. People who have a high degree of this trait, when exposed to extreme trauma (such as repeated severe child abuse, or natural disaster), may cope by dissociating their consciousness from the experience of the trauma. They achieve dissociation by entering a trance-like state. Where repeated trauma occurs, and dissociation is repeatedly used as an effective distress-reducing coping strategy, the process of dissociation is negatively reinforced. That is, the habit of dissociation is strengthened because it brings relief from distress. Eventually, sufficient experiences become dissociated to constitute a separate personality. These personalities may be activated at times of stress or trauma or through suggestion in hypnotic psychotherapeutic situations. Psychological treatment commonly involves helping the person integrate the multiple personalities into a single personality and develop non-dissociative strategies for dealing with stress. Unlike schizophrenia, psychotropic medication is of little value in treating the core symptoms of multiple personality disorder.

Multiple personality disorder is not classified with other personality disorders in ICD 10, nor is dissociative identity disorder classified with personality disorders in DSM IV. Rather they are classified with other dissociative conditions including dissociative amnesia (inability to recall events following a trauma); dissociative fugue (sudden unexpected travel away from the customary place of activities, coupled with confusion or amnesia for aspects of one's identity following a trauma); and depersonalization disorder (a sense of being out of one's body following a trauma). People who have a strong capacity to dissociate may develop one of the dissociative disorders rather than PTSD following trauma. Dissociative phenomena fall on a continuum, with tip-of-the-tongue experiences being a mild dissociative phenomenon, hypnotic trance states being a more pronounced dissociative condition, depersonalization disorder and dissociative amnesia and fugue being more extreme forms of dissociation, and dissociative identity disorder or multiple personality disorder being the most extreme form of dissociation. More extreme dissociative conditions develop when the person has a strong capacity to dissociate and is exposed repeatedly to severe trauma.

Schizophrenia is a complex disorder or group of disorders which affects 1 per cent of the population over 18 years of age. It has its onset in late adolescence or early adulthood. The condition is marked by positive symptoms such as delusions and hallucinations and negative symptoms such as inactivity and poverty of speech. While, in the past, a broad definition of schizophrenia was used in North America and a narrow definition was used in Europe, there is now considerable international agreement on a narrow-band definition of schizophrenia. The two syndrome hypothesis offers a way to integrate relevant findings of research into the biological origins of schizophrenia. This hypothesis entails the view that a distinction may be made between type 1 schizophrenia, which is a genetically inherited disease, marked by a dysregulation of the mesolymbic dopaminergic system and characterized by positive symptoms, and type 2 schizophrenia, which is a neurodevelopmental disorder arising from pre- or prenatal insults, marked by chronic negative symptoms.

Theories which posit specific cognitive deficits as an explanation for the range of schizophrenic symptomatology and which underpin cognitive behavioural approaches to treatment are well supported by a growing body of evidence from the laboratory and from treatment outcome studies. Social systems formulations which inform family systems approaches to therapy in schizophrenia are also supported by a considerable body of evidence from treatment outcome studies and field studies. Integrative diathesis-stress models of schizophrenia have also been developed which argue that, for the symptoms of schizophrenia to occur, a biologically vulnerable individual must be exposed to environmental stress. The interaction of the vulnerability factors with the stress factors leads to the occurrence of the symptomatology. Symptoms are subsequently maintained by ongoing exposure to environmental stress and by the way in which the person reacts to this stress and copes with the unusual experiences associated with schizophrenia. Treatment programmes premised on diathesis-stress models involve antipsychotic medication to address dysregulation of the dopamine system; family intervention to reduce family-based stress; and individual cognitive behavioural intervention to enhance personal coping strategies.

6

Personality Disorders

Often severe episodes of abnormal behaviour occur against a backdrop of more pervasive personality-based difficulties. To address this observation, distinctions are made in ICD 10 and DSM IV between episodic psychological disorders on the one hand and longstanding and persistent personality disorders on the other. Furthermore, it is accepted that in clinical practice many patients will present with both classes of disorders.

After considering the clinical features and epidemiology of personality disorders, a number of specific theoretical explanations for personality disorders will be considered. Earlier the broad theories of psychological problems on which these specific explanations are based are reviewed with reference to their main attributes, their contributions to our understanding and treatment of psychological problems, and their limitations.

Clinical Features

The defining features of personality disorders are:

- An enduring dysfunctional pattern of behaviour and experience
- The pattern begins in adolescence and is consistent across situations
- Difficulties with cognition, affect, impulse control,

behaviour and interpersonal functioning

- Recurrent relationship problems and/or occupational problems
- Difficulty learning from experience or benefiting from psychotherapy
- Associated with a history of other psychological disorders and criminality

Personality disorders are characterized by an enduring pattern of behaviour and experience that deviate markedly from cultural expectations and which lead to significant personal distress or significant impairment in social functioning. With personality disorders there are marked difficulties in two or more of the following domains: cognition, affect, impulse control, behaviour, and interpersonal functioning. With cognition, there may be peculiarities or difficulties in the way self, others and events are interpreted. At an affective level, the range, intensity, lability and appropriateness of emotional responses may be out of keeping with cultural expectations. There may be serious difficulties with impulse control, leading to highly erratic or impulsive behaviour, markedly inhibited behaviour, or peculiar behaviour.

With respect to interpersonal behaviour, there are typically serious difficulties making and maintaining stable and fulfiliing inter-personal relationships. Most people find the rigid behavioural patterns of people with personality disorders aversive and so avoid them. In the long term, the social isolation or negative response of others to people with personality disorders causes them personal distress. A hallmark of personality disorders is the fact that the cognitive, affective, behavioural and interpersonal difficulties constitute a longstanding and rigid pattern of psychological functioning. Usually personality disorders can be traced back to adolescence or early adulthood. Furthermore, individuals with personality disorders have great difficulty learning, from life experiences or therapeutic interventions, how to alter their rigid behaviour patterns. They repeatedly make the same mistakes and find it very challenging to learn from their errors. Often they are unaware of the impact of their behaviour on others, or wish to conceal their history of social and psychological difficulties. It is therefore useful, clinically, to include observations of family members in the assessment process. For some personality disorders, such as antisocial personality disorder, diagnosis is made by reputation rather than presentation. Personality disorders commonly occur in conjunction with other psychological difficulties or criminality. In clinics, personality-disordered patients create huge problems for staff and other patients. Staff describe them as manipulative and as playing one staff member off against another or as refusing to co-operate with treatment.

In DSM IV the ten main personality disorders are subdivided into three clusters on the basis of their cardinal clinical features. The first cluster includes the paranoid, schizoid and schizotypal personality disorders which are grouped together because they are characterized by odd or eccentric behaviour. The second cluster includes the antisocial, borderline, histrionic and narcissistic personality disorders which are characterized by dramatic, emotional or erratic-impulsive behaviour. The third cluster includes the avoidant, dependent and obsessive-compulsive personality disorders, all of which are characterized by anxiety and fearfulness. A very similar classification system is used in ICD 10. There are, however, some minor differences in classification. The schizotypal syndrome is listed as a psychotic condition along with schizophrenia in ICD 10 and narcissistic personality disorder is omitted. Also obsessive compulsive personality disorder is referred to as anakastic personality disorder. The clinical features of the ten main personality disorders are set out.

Multiple personality disorder (MPD), which was discussed earlier, is not classified with other personality disorders in ICD 10. Nor is dissociative identity disorder (the DSM IV term for MPD) classified with personality disorders in DSM IV. Rather, in both classification systems this condition is classified with other disorders where dissociation is the core feature.

Cluster A. The odd, eccentric group

The odd, eccentric group of personality disorders includes the following:

- Paranoid personality disorder
- Schizoid personality disorder
- Schizotypal personality disorder

Paranoid Personality Disorders

Clinical Features

People with paranoid personality disorder have a pervasive distrust of others. At a cognitive level, they interpret the motives of others in negative, malevolent, conspiratorial, or exploitative terms. They assume, on the basis of minimal or ambiguous evidence, that others intend to harm them, be disloyal to them, exploit them, or use personal information to discredit them. At an affective level, they are angry, combative and unforgiving of those whom they view as having harmed them. At a behavioural and interpersonal level, they constantly question the loyalty of close friends, partners or spouses. They may refuse to confide in others in case their confidences are used against them. They may constantly check on their spouses' or sexual partners' whereabouts and question their fidelity. They may hold grudges indefinitely if they believe that their friends or partners have harmed, insulted, injured or betrayed them.

Case Example

Margaret, aged 35, had a paranoid personality disorder. She and Brian came for therapy because of extreme marital distress. Brian complained that he felt like a prisoner in the marriage. Before leaving each morning Margaret interrogated him about his daily schedule.

Table.

Main clinical features of major personality disorders

Cluster	*Personality disorder*	*Main clinical feature*	*Cognition*	*Affect*	*Behaviour*	*Relationships*
Cluster A	Paranoid	Mistrust	Paranoid ideas	Aggressive	Secretive	Mistrustful
Odd, eccentric group	Schizoid	Detachment	Odd ideas	Constricted affect	Eccentric	Loner
	Schizotypal*	Eccentricity	Bizarre thoughts Ideas of reference Superstitious Strange speech	Constricted affect Socially anxious	Eccentric	Loner
Cluster B Dramatic, emotional, erratic group	Antisocial	Moral immaturity	No internalized set of moral standards	Aggressive	Violates rights of others Criminality	Multiple exploitative relationships
	Borderline	Impulsivity	Expectation of abandonment Shifts between over- and undervalued view of self and others	Impulsive Aggressive Depressed	Fights with others Self-harms	Multiple relationships with frantic unsuccessful attempts to avoid abandonment
	Histrionic	Seductive attention seeking	Belief in entitlement to attention no matter the cost to others	Shallow	Theatrical, seductive attention seeking	Multiple shallow relationships
Narcissistic**	Self-importance and grandiosity	Belief in entitlement to VIP treatment because they are better than others	Craves admiration and becomes angry or depressed if this need is frustrated	Admiration seeking	Multiple relationships which fail to meet their high expectations	
Cluster C Anxious, fearful group	Avoidant	Shyness	Belief that they will be rejected by others	Fear of rejection	Social withdrawal	Loners
Dependent	Lack of autonomy	Belief that they cannot function	Fear of autonomy	Refusal to take responsibility for decision	Multiple relationships in which they are completely	

(Cont..)

		autonomously		making Clinginess	dependent upon their partners
Obsessive-compulsive	Perfectionism	Belief that safety and security can be sustained through orderliness	Fear of imperfection	Detailed rule-following without regard for deadlines or overall goals	Multiple relationships in which their coldness and need for control cause conflict
* Schizotypal syndrome is listed as a psychotic condition with schizophrenia					

She phoned him frequently at work and would sometimes visit his office unexpectedly to check up on him. One night he found her checking through his wallet and the memory of his mobile phone to find clues about him having contact with another woman. Margaret complained that Brian had been unfaithful to her recently. He denied this but said that her suspiciousness was making infidelity an attractive option. Margaret's suspiciousness was a longstanding characteristic. She had a very small circle of friends, whom she had known since childhood. She would not make new friends because she found it hard to trust and confide in others. She also thought that new friends would ridicule her.

She had lost one of her closest female friends, Estelle, over an argument about loyalty. Estelle was ill on an occasion when Margaret had made arrangements for the two of them to go to a James Taylor concert. Margaret believed that she had gone to a party elsewhere on that evening. There was no evidence for this but Margaret believed that she had been betrayed. She would not forgive Estelle despite her attempts to put the incident to one side and continue the friendship. There were many incidents like this in Margaret's life, dating back to her childhood.

Margaret had grown up in a family where her parents had separated when she was 8 years old. In the time before the parents' separation she often heard her parents arguing about her father's whereabouts. These arguments would often end with her father storming out of the house and her mother shouting and crying. On some of these occasions she would say to Margaret, 'You can't trust anyone in this world.'

Margaret was attracted to Brian because of his openness and honesty and his willingness to spend a great deal of time with her. Brian found that after they were married the demands of work prevented him and Margaret from spending as much time together. As a result of this, Margaret began to accuse him of infidelity. He believed that if she continued to accuse him, despite his innocence, he would seriously consider separation. Margaret interpreted this as evidence of his infidelity.

Schizoid Personality Disorder

Clinical Features

. People with schizoid personality disorder show a pervasive pattern of detachment from social relationships. They are loners. At a cognitive level they have a preference for solitary activities and introspection. They have no desire for intimacy and do not wish to be involved with their families, friends or sexual partners. They are indifferent to praise or criticism. At an affective level they have a restricted range of emotions. They are not depressed, they do not find that relationships or activities bring them great pleasure, and they are not roused to anger even in extreme circumstances. At an interpersonal level they have few relationships and others find them cold and indifferent. They have little regard for social conventions.

Case Example

Norman, aged 20, had a schizoid personality disorder. He was a mathematics student who came for counselling because he was concerned that his addiction to an international multi-site computer game was interfering with his work. He stayed in a college hall of residence but lived an isolated life. This was the way he had always lived. As a kid he was ridiculed for being 'too brainy' and so became immersed in recreational mathematics. He refused to join a social skills group because he said he did not need to be involved with others.

Schizotypal Personality Disorder

Clinical Features

People with schizotypal personality disorder have unusual perceptual experiences, eccentric thoughts and speech, inappropriate or constricted affect, peculiar or eccentric behaviour, and a lack of close relationships. At a perceptual level, they may have unusual experiences. For example, they may sense the presence of ghosts or spirits, mystical forces or vibrations. They may experience depersonalization, that is, that they are an outside observer of themselves in a dreamlike state. They may also experience derealization, that is, that the world is dreamlike. At a cognitive level they may hold ideas of reference, that is, beliefs that routine events have a personal significance—for example, that the shape of a cloud means that they must carry out a particular activity. They may also hold paranormal convictions, about magic forces, telepathy, aliens and so forth, which are outside the norms of their subculture. They may also hold paranoid ideas and be suspicious of others. Their speech may be unusual, vague, eccentric or peculiar. At an affective level, people with schizotypal personality disorder show a constricted range of emotions and social anxiety, usually based on paranoid fears. At a behavioural level, people with schizotypal personality disorder are odd, eccentric and peculiar. For example, they may dress in an unfashionable and unkempt way and have unusual mannerisms. At an interpersonal level, people with schizotypal personality disorder are typically socially isolated and find making and maintaining friendships anxiety-provoking, so they avoid others.

Case Example

Silver, a man in his late forties, had a schizotypal personality disorder. He was a periodic outpatient at a psychiatric hospital. His appearance was distinctive. His long silver-grey hair and beard accounted for his unusual name. He wore a black overcoat, the hem of which trailed on the ground. He rarely washed and lived alone in a house that had belonged to his parents before they died. Silver had been referred to the psychiatric hospital originally by his family doctor years before, for participation in a group programme for people with schizophrenia. He refused to participate in the group programme because he was suspicious of other people, but requested periodic individual appointments with the psychologist who directed the group programme. In these outpatient individual sessions he insisted on discussing hypnosis and telepathy. He believed that he could hypnotize others from a distance; that he could see into the future; and that he could read minds. He had held these beliefs since his adolescence when he had numerous out-of-body experiences (depersonalization). Otherwise he was in contact with reality and was on no medication.

Cluster B. The dramatic, emotional, erratic group

The dramatic, emotional, erratic group of personality disorders includes the following:

- Antisocial personality disorder
- Borderline personality disorder
- Histrionic personality disorder
- Narcissistic personality disorder

Antisocial Personality Disorder

Clinical Features

People with antisocial personality disorder show a pervasive disregard for the rights of others and consistently violate these rights. This pattern has also been referred to as psychopathy, sociopathy, and dissocial personality disorder. At a behavioural level, people with antisocial personality disorder are consistently aggressive, destructive, deceitful and engage in theft and lying. Commonly in clinical practice, antisocial personality disorder is associated with a history of multiple arrests, multiple convictions and imprisonment. At a cognitive level, people with this personality disorder have not internalized rules for moral and ethical behaviour and are motivated by personal profit and pleasure seeking. They believe that aggressiveness, destructiveness, theft, deceit, and lying to achieve personal goals are justified. They develop elaborate rationalizations and cognitive distortions to justify their violations of others' rights. At an affective level, they are impulsive and reckless, displaying little planning and showing no anxiety even in the most dangerous situations.

The impulsivity and recklessness may find expression in an erratic occupational history, poor financial planning, speeding while driving, substance abuse and other forms of risk-taking which may lead to injury or death in some cases. People with antisocial personality disorder are aggressive and irritable and become involved repeatedly in fights. At an interpersonal level they have difficulty maintaining intimate relationships with friends or sexual partners. Commonly they charm people into believing that they wish to pursue a close relationship and then violate the trust the other person has placed in them through irresponsibility, disloyalty, aggression, destruction, theft or deceit. They show little remorse for violating trust in close relationships, and consequently have few if any close friends and typically a history of multiple sexual partners. The diagnosis of antisocial personality disorder is not given until after a person is 18, and commonly it is preceded by conduct disorder.

Case Example

Tony, who had an antisocial personality disorder, was referred to a psychiatric hospital from a prison, for psychological assessment. He was referred because he complained of depression to the visiting psychiatrist at the prison. He had a history of theft and occasional drug abuse. A thorough clinical interview and a full psychometric evaluation revealed no evidence of a mood disorder or, indeed, any other Axis I psychopathology. During the feedback session when the results of the assessment were presented to him, Tony said he had feigned depression because he wanted to be referred to the psychiatric hospital from the prison for a change of scene. He joked about the fact that he preferred the conditions in the hospital to the prison. During the psychological assessment he gave a history that drew a picture of himself as a stable, caring man who had fallen on hard times and so had stolen from time to time and unluckily been apprehended for occasional drug abuse.

They drew a picture of a man who had grown up in a disorganized family which his father left when Tony was a baby. His mother had a series of unreliable partners after that. Tony began rule-breaking and stealing as a child and had continued to do so right up to the present. He also truanted from school and began drinking and drug-taking in his teens. He had been married on four occasions and in each marriage had been violent towards his wife for trivial reasons. He had been involved in episodes of serious drug abuse and alcohol binges. Tony had been imprisoned on numerous occasions and had participated in a wide range of treatment programmes to help him alter his antisocial and drug-using behaviour patterns. All had been ineffective. He had no close friends, just transient acquaintances. His sister rarely saw him and his present wife (of a year's standing) was considering divorce.

Borderline Personality Disorder

Clinical Features

People with borderline personality disorder are highly impulsive and show a pattern of pervasive instability in interpersonal relationships, self-image and mood. At a cognitive level they have a core belief that, within the context of friendships, and relationships, they will be abandoned. At an affective level, they experience an intense fear of abandonment. They also have a propensity to experience intense uncontrolled aggression towards those whom they perceive as abandoning them. Their fear of abandonment leads to frantic, yet ineffective attempts to make and maintain relationships, particularly with sexual partners. Once in a relationship, their fear of abandonment may lead them to demand continued contact with their partner and continued caregiving and attention.

When their partners cannot allay their fear of abandonment or fail to meet their need for continued caregiving and attention, the person with borderline personality disorder may attempt to reduce the interpersonal distance by either becoming aggressive to their partner or making parasuicidal gestures. They may rage at their partners for frustrating their need for caregiving. They may then feel shame and guilt and engage in self-harm or self-mutilation to elicit continued caregiving from their partners. Thus they may cut themselves or take overdoses. Sometimes self-mutilation is carried out while in a dissociated state.

Within close personal relationships and therapeutic relationships, beliefs that people with borderline personality disorder hold about themselves and others alternate between extremes of idealization and devaluation. Thus they may view themselves or their partners as perfect, kind, caring and highly valued on one occasion, but on another (for example, when their need to receive care and attention from their partner is frustrated) view themselves or their partner as cruel, uncaring and despicable. This alternation between overvalued and undervalued views of the self and others and related changes in affect and behaviour may occur suddenly and unpredictably. Underlying the belief in the inevitability of abandonment, and the related fear and anger, is a pervasive sense of emptiness. People with borderline personality disorder may cope with this through impulsive spending, sexual activity, bingeing, drug use, reckless driving or other risk-taking activities. Commonly, borderline personality disorder occurs following a history of physical or sexual child abuse, neglect or early parental loss or death. A history of multiple partners and educational and occupational instability is common in cases of borderline personality disorder.

Case Example

Mary, aged 24, had a borderline personality disorder. She was referred for a parenting assessment to a child and family psychology service after her child was taken into foster care following a non-accidental injury. Mary had become frustrated with her 10-month-old daughter's continuous crying and bruised her badly by squeezing and shaking her. The parenting assessment showed that Mary had a good knowledge of how to care for her child but little sensitivity to the infant's signals; little understanding that the baby could not intentionally try to annoy her; and little tolerance for managing the routine daily demands of parenting. She also had a very limited social support network and difficulties making and maintaining friendships. Mary had been involved in several hetero-sexual relationships.

All had ended in flaming rows. In some instances she felt deep regret and a sense of being abandoned. Attempts to rekindle relationships with men always ended in major rows. She had a history of episodes of depression and had made a number of parasuicidal gestures. A general feeling of emptiness was occasionally broken by feelings of extreme joy (like at the start of a new relationship) and anger. She lacked any coherent life plan. Mary, herself, had been in care as a child on two occasions when her parents were unable to cope. She had also been regularly slapped and punched as a child when her father came home drunk. As a child, in school, she had never fitted in. She had left school at 14 and worked in a variety of casual jobs. She hoped when she met the baby's father, Kevin, that things would work out. But Kevin left her once she mentioned that she was pregnant. She hoped that her child and her role as a parent would give her happiness and a sense of direction. She was distraught when she found that it had brought her further misery.

Histrionic Personality Disorder

Clinical Features

People with histrionic personality disorder are characterized by a pervasive pattern of dramatic and excessively emotional attention-seeking. At a cognitive level there is the underlying belief that they are entitled to be the centre of attention and to have their needs for attention and admiration met, regardless of the cost to others. At an affective level they may feel a hunger for attention that they fear will not be fulfilled. At a behavioural level they routinely create situations in which they are the centre of attention. To do this they may express their views in dramatic and theatrical ways using excessive displays of emotion. However, when questioned in detail about why they hold particular views, typically they have little in the way of well-reasoned arguments or in-depth knowledge to back their assertions.

They are also highly suggestible and may change their views in response to changes in fads and fashions. This shallowness that characterizes their views and opinions also characterizes their relationships. They commonly consider their relationships with others (including health professionals) to be far more intimate and deeper than they are. So they may call doctors that they have seen infrequently by their first names and refer to acquaintances as their old friends. People with this personality disorder may dress and groom themselves so that they become the centre of attention. Often their self-presentation is deliberately seductive. Women with histrionic personality disorder may dress, make themselves up and behave in a way that is erotic, seductive and provocative. Men may dress and behave in a macho manner, drawing attention to their physical appearance, strength, athleticism and sexuality.

Case Example

Sarah, aged 18, had a histrionic personality disorder. She was originally referred to a child and family clinic in her final year at secondary school. She was to be excluded from school for instigating fights and disruption. She typically dressed like a film star, spoke like a soap opera character, and demanded everyone's attention. If she didn't get the attention she craved she threw a tantrum. She divided her favours between a number of boys at her school and incited them to compete with each other for her affections, promising each an exclusive relationship with her if they defeated the other boys. She was an only child and grew up in a family where her parents, who were involved in the arts, had little time for her. She spent much of her childhood alone and coped with the isolation by watching endless soap operas. She was intelligent but could not apply herself at school or later at college. She changed courses frequently and was led more by her attraction to partners and excitement than by vocational interests.

Narcissistic Personality Disorder

Clinical Features

People with narcissistic personality disorder have a pervasive pattern of grandiosity, a need for admiration, and a lack of empathy for others. At a cognitive level they believe that they are special or better than others and so are entitled to be treated differently. They have an over inflated view of their own accomplishments and believe that they are entitled to the best of everything and to associate only with people whom they perceive to be special or outstanding like themselves. They are preoccupied with fantasies of success, power and romantic love. At an affective level, they crave admiration, attention and deferential treatment from others. To confirm this view of themselves, at a behavioural level, they use their charm in social situations or their power in work situations to extract compliments and special treatment from others. Thus they may 'fish for compliments' in social situations and overload their employees in work situations to achieve their own personal goals. At an interpersonal level they have difficulty sustaining long-term relationships because they have a limited capacity for empathy and so do not appreciate the negative effect that their grandiosity and need for admiration have on others. They also leave many relationships because they fail to meet their high expectations of 'perfect love'.

They have difficulties maintaining peer friendships because they are arrogant, patronizing and envious of others' accomplishments and possessions and expect others to be envious of them. Occupationally, they may excel at their work or may avoid trying to excel for fear of failure. Furthermore, they are extremely sensitive to criticism, and when their partners or others frustrate their need for admiration or are the least bit critical, their self-esteem drops. They feel degraded, humiliated, hollow and empty. This may find expression as rage and anger or depression.

CASE EXAMPLE

Frank, aged 34, had a narcissistic personality disorder. He came for therapy at the insistence of his partner, Maria. She complained that he had become impossible to live with. Since they had met, Frank had always depended on her for emotional support and admiration. While she felt that this was appropriate when they were young lovers, she now felt that his demands were immature and excessive. He expected, for example, when he arrived home from work, for all household routines to cease and for Maria and the three children to devote their attention to Frank's accounts of his day's achievements. If Maria and the children did not meet his expectations he would fly into a rage and verbally abuse them. Later he would be remorseful and if he were not forgiven by Maria he would sink into a cold, detached, depressed state. The problem had become particularly bad recently when his business partner had questioned his judgement in a major business deal which went badly wrong and the company made a significant loss. Frank had relied on continued support and admiration from his business partner, Martin (whom he considered to be a lesser person than himself), and was devastated at his partner's criticism. He had oscillated between threatening to take legal action against Martin for defamation of character and withdrawing completely from frontline business transactions with the company's clients because of his incompetence.

Cluster C. The anxious, fearful group

The anxious, fearful group of personality disorders includes the following:

- Avoidant personality disorder
- Dependent personality disorder
- Obsessive-compulsive personality disorder

AVOIDANT PERSONALITY DISORDER

CLINICAL FEATURES

People with avoidant personality disorder show a pervasive pattern of social inhibition and shyness, beginning in adolescence or early adulthood. At a cognitive level, people with avoidant personality disorder believe that they are inferior to others, unlikeable, socially unskilled. They also believe that, when they meet new people, there is a high risk that they will be criticized, rejected, ridiculed, shamed or humiliated. At an affective level they experience intense anxiety in social situations. At a behavioural level they avoid situations, occupations, job promotions and pastimes that involve significant interpersonal contact with unfamiliar people. At an interpersonal level they have a constricted social lifestyle and avoid or show extreme restraint in intimate relationships, to avoid humiliation and embarrassment.

CASE EXAMPLE

Seamus, aged 45, was a bank official with avoidant personality disorder. He was referred to a communications consultancy centre for job interview preparation training. The process involved role-playing, videoing and reviewing the type of job interview in which he was due to participate to achieve promotion. While he was thoughtful and quite coherent during a conversation which preceded the role-playing, during the role-play he was virtually incoherent. He indicated after the interview that in the small rural branch of the bank where he worked he was just about able to tolerate the social anxiety he experienced when dealing with regular customers. He was shy and avoided all social contact outside work. He didn't want to be at our communications consultancy but the regional personnel manager of his bank said he needed to develop his job interview skills so he could be promoted to Assistant Manager level. He had attended our centre so as not to disappoint the personnel manager.

However, he would happily have avoided promotion because it would involve increased contact with unfamiliar people.

Dependent Personality Disorder

Clinical Features

People with dependent personality disorder show a pervasive pattern of submissiveness and clinginess. At a cognitive level, they have a belief that they must be taken care of by others (such as parents and partners) and that if separated from others their safety will be jeopardized. At an affective level, they experience extreme anxiety when separated from parents or partners whom they view as essential for their safety and security. At a behavioural level, they have difficulty making decisions without asking for and receiving advice and reassurance from others. They have difficulty disagreeing with others, particularly parents or partners, lest this lead to loss of support. They arrange for others to take responsibility for major areas of their lives and rarely initiate projects on their own. They go to great lengths to receive reassurance from others, even when this involves doing very unpleasant tasks. At an interpersonal level, when one intimate relationship ends, they quickly seek another, lest they be left to cope and make decisions alone.

Case Example

Tracy, aged 32, had a dependent personality disorder. She was the wife of a man who phoned requesting help with a sexual problem. He initially said that he was attracted to his 15-year-old daughter. It was suspected that sexual abuse had occurred in this case and so a full family assessment was offered. The family assessment revealed that he had had sexual intercourse with the daughter for over a year. We advised that the father leave the home and live separately while a programme of rehabilitation occurred. The programme would involve the mother and daughter strengthening their relationship, the daughter learning self-protection skills, and the father attending group therapy for sex-offenders. Tracy refused to co-operate because she felt unable to make decisions and function without her husband. Thus Tracy was prepared to jeopardize her daughter's safety for her own dependency needs.

Obsessive-compulsive Personality Disorder

Clinical Features

People with obsessive compulsive personality disorders show a pervasive pattern of preoccupation with orderliness, perfectionism, ethics, interpersonal control, and fiscal economy. At a cognitive level, there is a central belief that for safety and security to be maintained in all areas of life a carefully constructed set of organizational rules must be followed to perfectionistic standards. This belief covers the execution of routine activities of daily living; all areas of occupational responsibility; fairness in dealings with others; co-operating with others in jointly completing tasks; and managing money. At an emotional level, people with this personality disorder experience anxiety when their set of perfectionistic rules covering all areas of activity is not followed. At a behavioural level, a wide range of problems occur.

There is extreme difficulty finishing tasks on time, meeting deadlines, and in some instances tasks are never finished because perfectionistic standards are never reached. Flexible problem-solving also suffers, because a person with this personality disorder will become so fixed on following the original plan, that, even when unforeseen obstacles are encountered, the original plan is still followed. Leisure pursuits and family relationships receive little attention because of devotion to work-related tasks and this is not accounted for by economic necessity. If a person with obsessive compulsive personality disorder does involve themselves or their children in sports, their dedication to perfectionism makes the sporting events unhappy affairs marked by conflict. Practical solutions to problems at work or at home are difficult to find because of ethical scruples, beyond those accounted for by religious or cultural standards. In work, leisure and family situations, tasks are not delegated because of a concern that others will not do them to a perfectionistic standard. Conflict and eventually social isolation may arise at work or in leisure settings as a result of this. In economic matters, people with obsessive compulsive personality disorder are miserly and hoard money (however abundant) and possessions (however worthless) against possible future times of economic hardship.

At an interpersonal level, people with obsessive compulsive personality disorder may become isolated because they place such harsh demands on others to reach such high standards and because they have difficulty expressing tender feelings. This is because they can rarely find the perfect way to express their positive feelings for others and because they find it imperative to express their feelings of disappointment when others fail to meet their standards.

Case Example

Hank, aged 50, had an obsessive compulsive personality disorder. He was a divorced professor of experimental psychology. He came for therapy because of depression. This was related to the fact that his children and students had refused to have contact with him. In his home life he had always been meticulous and set the highest standards for family relationships and household routines. It was his anger when these standards were not reached that led to his divorce. However, he continued to have contact with his children. His insistence on punctuality and the critical attitude he took to his children's behaviour had led them to reduce the frequency of their visits with him and eventually to his wife suggesting that he have very infrequent access to the children. He was meticulous in his work. He wrote all his own computer programmes for conducting his experiments and analysing his data. He insisted that his dissertation and research students receive daily supervision and follow his guidance to the letter. If they did not meet his standards, he vilified them. His work was internationally known and all of his students had their work published or presented at major conferences. However, students who worked with him found his criticism of tiny errors in their work, his insistence that they follow his guidelines to the letter, and his excessive devotion to work to the exclusion of leisure difficult to take. They had complained to the dean about his criticism of them. He took the dean's suggestion that he take a more flexible approach and be less critical with his students very hard. He felt misunderstood and saw it as a personal attack.

Epidemiology and Classification

Personality disorders are classified in both DSM IV and ICD 10 as being distinct from episodic disorders such as anxiety or mood disorders. This classificatory distinction is achieved in DSM IV by placing personality disorders on a separate axis (Axis II) of a multi-axial system. The five axes of the DSM IV multiaxial system are:

- Axis I. Clinical disorders (e.g. depression)
- Axis II. Personality disorders (e.g. borderline personality disorder)
- Axis III. General medical conditions
- Axis IV. Psychosocial and environmental problems (e.g. occupational problems)
- Axis V. Global assessment of functioning (on a 100-point scale)

There is considerable evidence that patients with personality disorders are less responsive to treatment for Axis I disorders such as mood disorders, anxiety disorders, substance use disorders and eating disorders (Crits-Christoph, 1998; Roth and Fonagy, 1996). Estimates of the overall prevalence of personality disorders range from 10 to 13.5 per cent (Crits-Christoph, 1998). In Table below prevalence rates of personality disorders in community and clinical samples and gender differences in their distribution are presented. From the table it is clear that there is considerable variability in the prevalence of personality disorders in community samples, with rates ranging from 0.2 to 3.0 per cent depending upon the disorder. Personality disorders in cluster B, the dramatic-erratic group, are slightly more prevalent than those in cluster A, the odd-eccentric group, and these in turn are more prevalent than those in cluster C, the anxious-fearful group. The paranoid, schizotypal, antisocial, borderline, histrionic and dependent are the most common personality disorders in community samples. From Table below it may also be seen that there is considerable variability in the prevalence of personality disorders in clinical samples, with rates ranging from 1.0 to 31 per cent. In clinical samples the most common personality disorders are borderline, dependent, histrionic, schizotypal and avoidant. There are clear gender differences in the distribution of personality disorders and these are noted in Table also. All cluster A personality disorders are more common in males. In cluster B, antisocial and narcissistic personality disorders are more common in males, and borderline personality disorder is more common in females. In cluster C, obsessive-compulsive personality disorder is more common in males.

There is considerable comorbidity with personality disorders. Many people who have one personality disorder meet the criteria for a number of others. There is also considerable comorbidity of Axis I disorders with Axis II disorders. For example, 38-39 per cent of people with bulimia, 36-76 per cent of people with anxiety disorders, 36-65 per cent of people with mood disorders, and up to 75 per cent of people with substance abuse disorders have been found to have comorbid personality disorders (Crits-Cristoph, 1998).

Table.

Epidemiology of major personality disorders

Cluster	*Personality disorder*	*Prevalence in the community(%)*	*Prevalence in clinical samples (%)*	*Gender differences*
Cluster A	Paranoid	0.5-2.5	5.0	Higher in males
Odd, eccentric group	Schizoid	0.2-0.3	1.0	Higher in males
Schizotypal	2.5-3.0	17.0	Higher in males	
Cluster B	Antisocial	2.0	6.0	Higher in males
Dramatic, emotional, erratic group	Borderline	1.0-2.0	31.0	Higher in females
Histrionic	0.7-3.0	19.0	No difference	
Narcissistic	0.3-1.0	6.0	Higher in males	
Cluster C	Avoidant	0.3-1.0	16.0	No difference
Anxious, fearful group	Dependent	1.9-2.5	20.0	No difference
Obsessive-compulsive	1.0-1.9	8.0	Higher in males	
Note: Community prevalence rates and gender differnces are from DSM IV, Corbitt and Widiger (1995), Marmar (1988) and Weissman (1993). Clinical prevalence rates are based on				

The results of longitudinal studies, where the same group of patients is followed-up over time, and cross-sectional studies, where the prevalence of personality disorders in different age groups is examined, are fairly consistent. Overall, findings suggest that people with the disorders in clusters B and C show a reduction in symptoms over time, whereas those with cluster A disorders do not (Paris, 1996). It is now well documented that in midlife (the forties) people with antisocial and borderline personality disorder begin to mature out of their dysfunctional behaviour patterns. Borderline patients become less impulsive and self-harming and antisocial patients reduce their involvement in criminality. However, for both conditions only a minority establish successful relationships. About 10 per cent of borderline patients commit suicide.

There are marked similarities between each of the personality disorders and certain DSM IV Axis I disorders. These are listed in Table below. Paranoid, schizoid and schizotypal personality disorders each bear a marked resemblance to aspects of schizophrenia. However, people with these three personality disorders do not show frank delusions, hallucinations, thought disorder, or negative symptoms. There are clear parallels between antisocial personality disorder and conduct disorder since both involve moral immaturity and violation of others' rights. About a third of youngsters with conduct disorder in adulthood meet the diagnostic criteria for antisocial personality disorder (Kazdin, 1995). Borderline personality disorder resembles depression, in so far as loss and abandonment are central themes and negative affect is a core feature of both conditions. Histrionic personality disorder and conversion hysteria are similar in so far as in both instances a caricature of a role is adopted and this is associated with attention-seeking and secondary gain.

However, with hysterical personality disorder, the role is that of a stereotypically seductive woman or macho male, whereas in conversion hysteria a sick-role is adopted. The symptoms of conversion hysteria, which are not feigned, include deficits in sensory or motor functioning in the absence of an organic illness. Avoidant personality disorder and social phobia are similar in so far as shyness typifies both conditions. However, those with the personality disorder avoid relationships, whereas those with the phobia avoid particular social situations. Dependent personality disorder and separation anxiety both entail problems with autonomy. Separation anxiety is a childhood disorder, but dependent personality disorder is the adult expression of the same core issues. Obsessive-compulsive personality disorder closely resembles obsessive-compulsive disorder. However, with the personality disorder the obsession with orderliness and rule-following is fully accepted and embraced as a valued part of the person's lifestyle, whereas, with obsessive-compulsive disorder, the obsessions and compulsions are resisted.

Table.

Comparison of personality disorders and Axis 1 disorders with similar features

Cluster	*Personality disorder*	*Similar Axis 1 disorder*	*Common clinical features*	*Differences in clinical features*
Cluster A Odd, eccentric group	Paranoid (PPD)	Schizophrenia	Mistrust and suspiciousness	In PPD, delusions, hallucinations, thought disorder, and negative symptoms of schizophrenia are absent
				In SDPD, delusions,

(Cont..)

Schizoid (SDPD)	Schizophrenia	Attachment problems and social isolation	hallucinations, thought disorder, and negative symptoms of schizophrenia are absent
Schizotypal (SLPD)	Schizophrenia	Eccentric thoughts, perceptual experiences and speech	In SLPD, delusions, hallucinations, thought disorder, and negative symptoms of

ETIOLOGICAL THEORIES

Diathesis-stress, psychodynamic, and cognitive-behavioural theories of personality disorders have been developed and summaries of these will be outlined below (Clarkin and Lenzenweger, 1996). However, before proceeding to these, an alternative way of conceptualizing enduring patterns of dysfunctional behaviour and experience deserves mention. This alternative is trait theory.

TRAIT THEORIES

Within DSM IV and ICD 10, personality disorders are conceptualized in categorical terms. That is, it is assumed that within a population some people have personality disorders and some do not and that there are qualitative differences between those that do and do not meet the diagnostic criteria for personality disorders. Trait theories, in contrast, argue that a limited number of dimensions may be used to characterize important aspects of behaviour and experience. Traits are normally distributed within the population. So for any given trait (for example, introversion-extroversion) most people show a moderate level of the trait, but a few people show an extremely low or extremely high level of the trait. Within a population, people who fall at the extreme ends of these dimensions may have the sorts of difficulties attributed in DSM IV and ICD 10 to people with personality disorders. However, these people differ from others only in the degree to which they show particular traits.

Within the broad tradition of trait theory there is considerable controversy over the precise number of traits that may appropriately be used to describe personality functioning. For example, Eysenck (1990) argued that three traits (neuroticism, introversion and psychoticism) could account for most aspects of personality functioning. In contrast, Cattell (1990) argued that sixteen traits were required. The arguments of Eysenck and Cattell were based in part on the results of factor analyses of responses of large numbers of people to questionnaires designed to measure all aspects of personality functioning. Factor analysis is a procedure for mathematically clustering questionnaire items that correlate with each other together into a small set of dimensions. There are many different types of factor analysis, and the different numbers of dimensions in Eysenck's and Cattell's trait theories of personality reflect, in part, the different types of factor analysis they used. It is now accepted that the sixteen narrow-band traits in Cattell's model can be combined to produce the three broad-band traits in Eysenck's model. Eysenck showed that his three dimensions of personality are highly heritable and reflect under-lying individual differences in neurophysiological functioning. Introversion reflects differences in physiological arousal level.

Introverts have a high level of arousal so they limit interpersonal contacts, while extroverts have a low level and so seek interpersonal contact. Neuroticism reflects differences in emotional liability at a psychological and physiological level, with high scorers being extremely labile. Psychoticism reflects interpersonal insensitivity and high scores on this dimension are associated with a disregard for social conventions and solitariness. Eysenck (1987) has argued that cluster A personality disorders (the eccentric group) are characterized by high levels of introversion; cluster B personality disorders (the dramatic-erratic group) are characterized by high levels of psychoticism; and cluster C personality disorders (the anxious group) are characterized by high levels of neuroticism.

In recent years trait theory has come to be dominated, particularly in the USA, by the Five Factor Model of Personality, which builds on the earlier insights of Eysenck, Cattell and others (Costa and Widiger, 1994). This model includes the following dimensions: extroversion, neuroticism, openness to experience, agreeableness, and conscientiousness. The first two dimensions are the same as those proposed by Eysenck. Furthermore, the traits agreeableness and conscientiousness are two aspects of Eysenck's psychoticism factor. Disagreeable people are interpersonally cold and people low on conscientiousness disregard social conventions. Openness to experience refers to a dimension that extends from imaginative creativeness to constriction.

The profiles of major personality disorders on the dimensions of the Five Factor Model of Personality are presented in Table below.There is good evidence for the heritability of four of the five factors of the Five Factor Model (Costa and Widiger, 1994). Agreeableness is more environmentally determined than the other four factors. Trait theory has few therapeutic implications, but it does offer a more reliable and valid approach for assessment. Many of the personality questionnaires that have been developed to measure personality traits have good reliability (Pervin, 1990). With respect to validity, trait models fit with the observation that traits are normally distributed within the population.

Table.

The profiles of personality disorders on the Five Factor Model of Personality

Cluster	Personality disorder	Neurotidsm	Extroversion	Openness	Agreeableness	Conscientious ness
High		Anxious	Sociable	Creative	Compassionate	Reliable
Low		Calm	Reserved	Conventional	Cynical	Negligent
Cluster A						
Odd, eccentric group	Paranoid		low	low	low	
Schizoid		low				
Schizotypal	high	low	high			
Cluster B						
Dramatic, emotional, erratic group	Antisocial	low			low	low
Borderline	high	high		low	low	
Histrionic	high	high	high		low	
Narcissistic	high	high		low	high	
Cluster C						
Anxious, fearful group	Avoidant	high	low			
Dependent	high	high		high		
Obsessive-compulsive	high	low	low		high	
Note: Adapted from Widiger (1993). Bold face indicates that extreme scores on the dimension						

Compared with categorical classification systems, trait models offer a more parsimonious way of describing people with rigid dysfunctional behaviour patterns.

DIATHESIS-STRESS THEORIES

Diathesis-stress theories of personality disorders argue that both biological factors and stressful environmental factors, particularly those within the individual's family of origin, contribute to the development of personality disorders. These theories entail the view that people with certain genetically determined temperamental characteristics develop particular personality traits and that personality disorders emerge when such people are exposed to certain types of psychosocial risk factors within their families or wider social systems (e.g., Paris, 1996). The weight of evidence shows that 50 per cent of the variance in major personality traits such as extroversion, neuroticism, openness to experience, and conscientiousness may be accounted for by genetic factors (Paris, 1996; Crits-Cristoph, 1998).

The mechanisms by which genetic factors influence personality traits are complex. Probably multiple genes determine temperamental characteristics, and these interact with environmental influences in the development of personality traits. There is considerable evidence from longitudinal studies of the link between temperament and personality traits. Children with high activity levels and positive affect become extroverted. Children who are highly irritable and fearful show high levels of neuroticism in later life. Children who show attentional persistence later develop high levels of conscientiousness (Rothbart and Ahadi, 1994). Children with extreme temperamental characteristics may be more vulnerable to environmental stressors, or they may elicit reactions from parents and others that exacerbate their extreme temperamental characteristics.

In contrast to the evidence for the role of genetic factors in the development of personality traits, there is little evidence for a major role of genetic factors in the development of personality disorders. Thus, it is probable that people with extreme levels of particular personality traits (which are 50 per cent heritable), when exposed to particular types of family environments, develop certain personality disorders. In prospective and retrospective studies, a wide variety of family-based risk factors have been found to predispose people to the development of personality disorder. These include: separation from or loss of a parent; parental psychopathology and related impaired parenting; problematic parent-child relationships; extremely low or high levels of family cohesiomn; physical and sexual abuse; neglect; and the absence of social support (Paris, 1996).

Evidence for the specific role of biological or genetic factors in the etiology of specific personality disorders deserves mention. Within cluster A, there is some evidence that biological factors may play a role in the development of schizotypal personality disorder. Schizotypal personality disorder may form part of a spectrum of psychotic or psychotic-like disorders that includes schizophrenia. People with schizotypal personality disorder have been found to have a positive family history for schizophrenia; a dysregulation of the dopamine system similar to that found in people with schizophrenia; and many of the attentional and information processing deficits found in people with schizophrenia (Siever and Davis, 1991).

Within cluster B, there is some evidence for the role of biological factors in the development of antisocial personality disorder. People with antisocial personality disorders have very low levels of physiological arousal and fail to develop conditioned responses to fear-related stimuli (Mednick and Moffit, 1985). Thus they are unable to learn from negative experiences. This lack of conditionability reflects a neuropsychological vulnerability to developing antisocial behaviour. However, this vulnerability only leads to antisocial personality disorder when combined with psychosocial risk factors, particularly family disorganization, paternal criminality, maternal psychological difficulties, parental alcoholism, and inconsistent discipline (Kazdin, 1995). The evidence on the role of genetic factors in borderline personality disorder is conflicting. However, it has consistently been found that people with borderline personality disorder have a history of child abuse, particularly child sexual abuse (Paris, 1996).

Within cluster C, an anxious temperament and the personality trait neuroticism (both of which are approximately 50 per cent heritable) predispose individuals to developing avoidant and dependent personality disorders. However, people with these attributes require exposure to a highly enmeshed family culture in which there is a high degree of parental control and little encouragement for the development of autonomy for the development of avoidant and dependent personality disorders (Paris, 1996).

Psychopharmacological treatments for personality disorders are not well developed; the current state of such treatments has been reviewed recently by Woo-Ming and Siever (1998). There is some evidence that antipsychotic agents may lead to partial symptomatic relief for cluster A personality disorders, particularly schizotypal personality disorder. For cluster B personality disorders, particularly borderline personality disorder, antidepressants have been shown to improve mood regulation, and fluoxetine, lithium and carbamezapine have been shown to reduce impulsivity in some trials. For cluster C personality disorders, particularly avoidant personality disorder, MAOI antidepressants have been shown to decrease symptomatology.

Marital and family therapy interventions for personality disorders focus on helping couples and families recognize and alter current family-based patterns of interaction that maintain the personality disorder. They also help family members identify and challenge the belief-systems and narratives that underpin these rigid interaction patterns. Finally, in marital and family therapy, family members may be invited to understand how particular personal characteristics or traits and particular family-of-origin experiences have predisposed them to developing family belief systems and behaviour patterns which maintain personality disorders. Controlled trials of marital and family therapy for personality disorder have not been reported in the literature, but guidelines on current best practice have been published (e.g., Sperry, 1999).

Psychoanalytic Theories

In modern psychoanalytic practice, Kernberg's (1996) approach to understanding and treating borderline personality disorder has been particularly influential. Kernberg's psychoanalytic object relations theory assumes that, during the course of early development, individuals evolve through a series of stages in which their internal representations of self and others become increasingly sophisticated. In object relations theory, it is assumed that the child learns about others through their relationship with their mother or caregiver. 'Object' is the technical term used in object relations theory to refer to others (as distinct from the self) and particularly to the mother during early development. At a very early stage of development, the infant does not distinguish between representations of the self and the mother.

The main distinction that is made is between experiences that make the child feel good or bad. So at the earliest stage of development the child develops symbiotic fused self-object representations that are 'all good' or 'all bad'. At a second stage of development, the child learns that the self and the mother (or object) are separate. At this stage the child develops representations for an 'all-good self and an 'all-bad self'; an 'all-good object' and an 'all-bad object'. As the child matures into the third stage of development, the 'all-good' and 'all-bad' self-representations are integrated. The child develops a more complete view of the self as having both positive and negative impulses and wishes. Concurrently the 'all-good' and 'all-bad' representations of others (parents, siblings, friends) are integrated. The child develops representations of others as having both positive and negative attributes. When this happens the child becomes capable of having realistic relationships in which ambivalent feelings towards others can be tolerated. So the child may feel that they love their parents who are good a lot of the time and a bit annoying some of the time.

Where parents are overindulgent or overly neglectful, aggressive or controlling, the child fails to develop mature self-other object relations. In their relationships with others splitting occurs. They view others as 'all-good' idealized rescuers who will meet all their needs or 'all-bad' persecutors who are out to harm them. They also oscillate between viewing the self as 'all-good' or 'all-bad'. These difficulties are the hallmark of all personality disorders. Different types of personality disorders develop depending upon the person's temperament, whether the child was overindulged or neglected, the degree to which this occurred, and the specific defence mechanisms that they used to cope with forbidden sexual and aggressive impulses. Cluster A and B personality disorders reflect the outcome of lack of care, neglect, rejection or abuse; whereas cluster C reflects the outcome of excessive parental control or overprotection.

Defences are psychological strategies used to cope with conflict between unacceptable impulses (often unconscious sexual or aggressive urges from the id) and the prohibitions of the conscience (or superego). Thus, if a person experiences an unacceptable impulse, anxiety about the consequences of acting on this impulse will be experienced. Defences are used to reduce anxiety. Defences are essential, but some are more adaptive than others. From Table below it may be seen that the defences of splitting and projection, which typify many personality disorders, particularly borderline disorder, are at the most primitive level. Splitting involves reverting to viewing the self and others in 'all-good' or 'all-bad' terms. Projection involves attributing negative aspects of the self to others.

Psychodynamic psychotherapy has been found in a small number of trials to improve adjustment in patients with borderline personality disorder, and other types of personality disorder also (Stevenson and Meares, 1992; Munroe-Blum and Marziali, 1995; Winston *et al.*, 1994). Psychodynamic psychotherapy first involves identifying and interpreting the part-object relationship in the transference and the alternating representations of the self and object within this part-object relationship.

Table.

Defence mechanisms at differnent levels of maturity

Maturity level	*Defence*	*The individual regulates emotional discomfort associated with conflicting wishes by...*
Mature defences	Self-observation	dealing with conflict by monitoring how situations lead to conflicting wishes or negative affect and using this new understanding to modify negative affect
	Humour	reframing the situation that gives rise to conflict or negative affect in an ironic or amusing way
	Self-assertion	expressing conflict-related thoughts or feelings in a direct yet non-coercive way
	Sublimation	channelling negative emotions arising from conflicting wishes into socially acceptable activities such as work or sports
Neurotic defences	Repression	expelling unwanted thoughts, emotions or wishes from

(Cont..)

Denial	refusing to acknowledge the painful features of their situation or experi ences which are apparent to others
Reaction formation	Substituting acceptable behaviours, thoughts or feelings which are the opposite of unacceptable or unwanted behaviours, thoughts or feelings that arise from a conflict
Displacement	transferring negative feelings about one person onto another, less threatening person
Rationalization	providing an elaborate self-serving or self-justify ing explanation for unacceptable thoughts, actions or feelings that arise from a conflict

Table. *(Continued)*

Maturity level	*Defence*	*The individual regulates emotional discomfort associated with conflicting wishes by…*
Immature defences	Splitting	failing to integrate the positive and negative qualities of self and others and viewing self and others as either

(Cont..)

	all good or all bad
Projection	attributing to others one's own unacceptable thoughts, feelings and wishes. Projection typically occurs in conjunction with splitting
Passive aggression	unassertively expressing aggression towards others in authority by overtly complying with their wishes while covertly resisting them

Note: Based on Conte and Plutchik (1995).

For example, the therapist may point out that in one instance the therapist-client transference relationship resembles that of a persecuting parent to a frightened and needy child, with the client in the role of the child, and that at another time the relationship is the same but the roles are reversed, with the client adopting the role of the persecuting parent. Psychodynamic psychotherapy secondly involves interpretations of the links between such negative part-object relationships activated in the transference at some times, and positive or idealized part-object relationships activated in the transference at other times. For example, the therapist may point out that, while the persecuting parent-needy child roles typify the transference relationship at one time, at others the therapist-client relationship is like that between a satisfied child and an all-giving mother.

The third step in psychodynamic psycho-therapy involves interpreting the links between the use of splitting, primitive idealization and projection as defences to reduce anxiety associated with attempting to integrate the all-good and all-bad primitive object relations. For example, the therapist may point out that one reason for viewing people as all-good or all-bad is that it preserves the possibility of having one's needs met by an all-gratifying mother. However, the down-side of using splitting and projection is that it prevents the development of sustainable intimate relationships because it requires denying the existence of frustrating characteristics in people defined as all-good, and positive characteristics on people defined as all-bad. Through these types of interpretation, psychodynamic psychotherapy facilitates the integration of dissociated all-good and all-bad primitive object relations and related polarized affects, which in turn leads to improved affect and impulse control. Kernberg *et al.* (1989) have documented their treatment techniques in a therapy manual for working with borderline personality disorder.

Cognitive-behavioural Theories

Cognitive-behavioural theories of personality disorders, of which Beck's is the prime exemplar, argue that people with personality disorders have developed pervasive, self-perpetuating cognitive-interpersonal cycles that are severely dysfunctional (Pretzer and Beck, 1996). Early life experiences, including family routines and relationships as well as traumatic events, lead to the formation of assumptions about the world and in particular about inter-personal relationships—for example, 'people are not trustworthy'. In day-to-day interactions, these underlying assumptions lead to automatic thoughts such as 'he's trying to con me'. This in turn leads to emotional reactions such as anger, and behavioural reactions such as oppositional and confrontative conversation. This in turn elicits behaviour from others such as secretiveness and avoidance which reinforce the basic assumption that 'people are not trustworthy'. For each personality disorder, there are predominant mood states and predominant behavioural strategies used to deal with interpersonal situations. Collections of basic assumptions, learned in early life, may be formed into schemas which inform those aspects of the world to which the person attends and how they are apt to interpret most situations.

For example, an abuse-mistrust schema may contain a collection of beliefs about the untrustworthiness of others and their potential for abusing or harming the person. In addition to schemas, cognitive distortions such as mind-reading ('I just know he's trying to get at me with that remark') or emotional reasoning ('I feel angry, so he must be persecuting me') contribute to the way a person reacts to interpersonal situations. Also, predominant mood states may predispose people to attend to particular types of information or to evaluate situations in particular ways. For example, anger may predispose a person to attend to potential threats and to evaluate situations as opportunities for confrontation.

Cognitive therapy aims to break the dysfunctional cognitive-interpersonal cycles which constitute the person's personality disorder, using a variety of cognitive and behavioural strategies. These include: helping people learn to identify and challenge their automatic thoughts and core assumptions; helping people develop different interpersonal strategies and skills which are less likely to elicit from others behaviour that reinforces dysfunctional beliefs; and helping people engage in activities that will directly alter their mood states.

A couple of controlled trials of therapies developed within the cognitive behavioural tradition have been conducted (Linehan *et al.*, 1991; Alden, 1989). Linehan *et al.* (1991) found that women with borderline personality disorder treated with dialectical behaviour therapy engaged in fewer parasuicidal behaviours and had fewer days of hospitalization compared with cases who received routine outpatient treatment. Also, the attrition rate of 17 per cent for cases receiving dialectical behaviour therapy was much lower than the attrition rate of 58 per cent shown by the control group.

Dialectical behaviour therapy (Linehan, 1993) is a complex multisystemic intervention programme, which includes intensive individual and group work with the patient for approximately eighty sessions over a period of a year, and carefully planned case management work with significant members of the patient's social network, including family members, friends and other involved professionals. The overall programme is goal-focused and goals include: decreasing suicidal behaviour; decreasing behaviours that interfere with engaging in therapy and maintaining a good quality of life; decreasing PTSD symptomatology; increasing behavioural skills; and increasing self-respect. Group therapy includes psycho-education, social skills training, and coaching in impulse control and affect regulation. The central feature of individual work in dialectical behaviour therapy involves using dialectical strategies to hold both sides or polarities expressed by clients in focus for a sufficiently sustained period for clients to achieve a more integrated and flexible position.

Alden (1989), in a controlled treatment outcome study of patients with avoidant personality disorder, found that ten sessions of closed-group behavioural treatment led to significant improvements in patient adjustment compared with a no-treatment control group. In this study three distinct behavioural treatments were compared: graded exposure, standard social skills training, and intimacy-focused social skills training. All were found to be equally effective. Similar results have been found in uncontrolled studies of behavioural treatments (Crits-Cristoph, 1998).

Personality disorders are characterized by enduring dysfunctional patterns of behaviour and experience which begin in adolescence and are consistent across situations. There are difficulties with cognition, affect, impulse control, behaviour and interpersonal functioning. There are also recurrent relationship problems or occupational problems, with a history of other psychological disorders or criminality.

People with personality disorders have difficulty learning from experience or benefiting from psycho-therapy. In DSM IV, ten main personality disorders are subdivided into three clusters on the basis of their cardinal clinical features. The odd, eccentric cluster includes the paranoid, schizoid and schizotypal personality disorders. The dramatic, emotional, erratic cluster includes the antisocial, borderline, histrionic and narcissistic personality disorders. The third cluster includes the avoidant, dependent and obsessive-compulsive personality disorders, all of which are characterized by anxiety and fearfulness.

A very similar classification system is used in ICD 10. The prevalence of personality disorders is approximately 10-13 per cent. The long-term outcome is better for both the dramatic, erratic and the anxious, fearful clusters than for the eccentric cluster. Trait conceptualization of variations in personality, such as the Five Factor Model of Personality, may be more parsimonious than categorical typologies such as those contained in DSM IV and ICD 10. About 50 per cent of the variance in personality traits may be accounted for by genetic factors. According to diathesis-stress theories, when people with extreme levels of personality traits are exposed to particular family-based risk factors, personality disorders develop. Diathesis-stress theories argue that personality disorders emerge when genetically or constitutionally vulnerable individuals are exposed to particular types of environmental stresses. These theories may inform both psychopharmacologically-based and family-based treatment programmes. Psychodynamic and cognitive behavioural theories of personality disorders have also been developed. Controlled trials of treatment programmes based on these conceptualizations have shown them to be effective for some personality disorders.

7

Models of Abnormal Behaviour

Various clinical problems described earlier brings out four main categories of explanations have been offered reflecting four broad models of abnormal behaviour:

- Biological model
- Psychoanalytic model
- Cognitive-behavioural model
- Family systems model

In this chapter the main assumptions of each of the models will be outlined along with an account of each model's achievements or strengths and shortcomings or limitations. A summary of these issues is given in Table other page.

Biological Model

Assumptions

The biological model of abnormal behaviour is also referred to as the organic model, the traditional medical model, or the disease model (Tyrer and Steinberg, 1998). In this model it is assumed that the various abnormal behaviours that people with a particular syndrome display are symptoms of a specific disease with a discrete cause, a unique course and prognosis, and for which a specific physical treatment will ultimately be identified. This model evolved within a medical tradition where there were numerous examples of physical conditions involving a syndrome of signs and symptoms that could be explained by a discrete cause, such as an infection or metabolic dysfunction. Syphilis is a good example of a condition where a discrete physical cause (syphilitic infection) causes a psychological syndrome (general paresis of the insane) and can be treated by specific physical methods (inoculation). In 1897 Richard von Krafft-Ebing (1840-1902), a German neurologist, following the work of Louis Pasteur (1822-1895), inoculated patients with general paresis, with pus from syphilitic sores, and this halted the development of the degenerative condition which typically culminated in insanity.

The success of this work gave impetus to the biological model of psychiatry. Not surprisingly, the biological model has been championed by psychiatry more than other mental health professions. The diathesis-stress model is a subtler variant of the biological model. In this type of model it is assumed that psychiatric illnesses or psychological problems occur when people who are biologically vulnerable to such difficulties are exposed to particular stresses.

Achievements

The biological or medical model has led to a number of important achievements. The first of these has been the development of mental health legislation. This legislation makes provision for the involuntary detention and/or treatment of people with psychological disorders whose judgement is severely impaired, especially if they are a danger to themselves or other people. Suicidal behaviour and extremely violent behaviour are recognized within mental health legislation in most countries as possible reflections of impaired judgement and possible grounds for involuntary detention or treatment in a psychiatric treatment centre.

A second important achievement of the biological or medical model has been the development of widely used classification systems. This began with the painstaking work of Kraepelin (1896) who, through careful observation, catalogued symptoms characteristic of different groups of patients. Psychological problems are currently classified in the World Health Organization's (1992) *The International Classification of Diseases—Tenth.*

Table.

Model of abnormal behaviour

	Biological model	*Psychoanalytic model*	*Cognitive behavioural model*	*Family systems model*
Cause of abnormal behaviour	Central nervous system abnormality	Unconscious psychopathology	Learned habits	Dysfunctional family system
Therapy goal	• Rectify CNS abnormality	• Resolve unconscious conflicts	• Learn more adaptive habits	• Alter problem maintaining family interaction patterns and beliefs
Therapy process	• Pharmacological • ECT • Psychosurgery	• Long-term one-toone psychoanalysis	• Short-term cognitive behaviour therapy	• Short-term marital or family therapy
Achievements	• Liberation of insane • Mental health legislation • Classification • Scientific method • Medical technology	• Discovery of unconscious • Makes madness meaningful • Discovery of transference • Legitimized talking cure • Inspired other	• Brief and effective • Permits evidencebased practice • Has specific interventions for specific problems	• Highlighted social context of abnormal behaviour • Brief and cost-effective • Permits evidence-based practice • Permits

(Cont..)

	• Pharmacological and physical	models of personality and therapy	• Scientifically rigorous	integration of biological, psychological
• Gives framework for managing complex, multiproblem families				
Limitations	• Discounts environmental influences • Discounts dimensional models • Side-effects of medications • Discounts sociopolitical factors • Promotes exclusion • Mental health legislation is abused • Marginalizes non-medics	• Contains untestable hypotheses • Inaccurate account of infantile sexuality • Not cost-effective	• Danger of trivializing problems • Danger of discounting organic factors • Danger of discounting social system factors	• Danger of vagueness • Danger of discounting organic factors • Danger of discounting intrapsychic factors

Edition: Classification of Mental and Behavioural Disorders (ICD10) and in the American Psychiatric Association's (1994) *Diagnostic and Statistical Manual of Mental Disorders* (now in its fourth edition). These two systems—ICD 10 and DSM IV—are now used widely throughout the world and provide a way for clinicians and researchers to communicate with each other in a relatively unambiguous manner. This was not always the case and in the past there has been wide variation in diagnostic criteria used in different countries. For example, in the US a very broad definition of schizophrenia was typically used up until the development of DSM III in 1980. The extraordinary differences in rates of diagnosis are highlighted by the results of the US-UK project presented in Table below.

The use of scientific methods and quantitative techniques to study abnormal behaviour is a third achievement of the biological model. Case-control studies, longitudinal studies, and randomized controlled trials are three common research designs used in studies conducted within the medical tradition.

Case-control studies may be used to determine the unique characteristics associated with a particular psychological disorder. In case-control studies a group of diagnostically homogeneous cases, about which the researcher is trying to find out more information, is compared to another group of patients with a known condition or to a normal control group. This was the method Kolvin *et al.* (1971) used to differentiate between childhood schizophrenia and autism. He found that, compared with childhood schizophrenia, autism was associated with an earlier onset, delayed language development, catastrophic reactions to environmental changes, gross stereotypic behaviours, and the absence of delusions and hallucinations.

Table .

Rates of diagnosis of schizophrenia in US and UK

	Hospital diagnosis	*Project diagnosis*
US	61%	29%
UK	34%	35%

Note: From Cooper *et al.* (1972). N=192 in US. N=174 in UK

Longitudinal studies may be used to determine the course of a psychological disorder. In longitudinal studies the researcher assesses a group of diagnostically homogeneous cases at repeated time points. This was the research design used by Robins (1966) to establish the course of conduct disorder. He found that about a third of cases developed antisocial personality disorder and a third adjusted well in adulthood. The remaining third had a more variable course.

Randomized controlled trials are used to evaluate the effectiveness of specific treatments. Diagnostically homogeneous cases are randomly assigned to treatment or control conditions. All cases are assessed immediately before and after treatment and then a number of months after treatment has been completed. Standard assessment procedures are used for all cases at all three time points. If the group that receives treatment shows greater improvement than the control group, then the treatment may be said to be effective. This design is typical of studies reviewed in books such as *A Guide to Treatments that Work* (Nathan and Gorman, 1998) and *What Works with Children and Adolescents?* (Carr, 2000a).

The development of sophisticated methods for monitoring activity within the central nervous system is a fourth achievement of the biological model. These include brain imaging systems such as the CAT (computerized axial tomography) scan, the PET (positron emission tomography) scan, the SPECT (single photon emission tomography) scan, and fMRI (functional magnetic resonance imaging). Imaging techniques have thrown light on those areas of the brain associated with particular disorders. For example, studies using SPECT have found increased metabolic rates in the basal ganglia in people with obsessive-compulsive disorder (Edmondstone *et al.*, 1994).

A fifth achievement of the biological model is the development of physical treatments for psychological problems. These include psychopharmacological advances such as neuroleptic medication for psychoses; antidepressant medication for unipolar mood disorders; and lithium carbonate for bipolar disorder (Nathan and Gorman, 1998). Electroconvulsive therapy is also an effective physical treatment for major depression which does not respond to medication and psychotherapy (Lock, 1999). Psychosurgery has been found to be effective in extreme cases of OCD. The development of physical treatments, particularly psychopharmacological treatments, has depended upon hypotheses in which dysregulation of neurotransmitter systems has played a central part—for example, the dopamine theory of schizophrenia (Kahn and Davidson, 1995).

LIMITATIONS

Despite these major achievements, the biological model is not without its limitations. First, much abnormal behaviour is not caused exclusively by organic factors. Rather, for conditions such as anxiety, depression and schizophrenia, people with a genetic vulnerability to a particular condition develop abnormal behaviour if they are exposed to particular stresses within their environment. This observation has led people working within this tradition to move away from a purely biological model to a diathesis-stress model of psychological problems. Such models acknowledge the importance of psychosocial factors such as family environment, gender, social class and culture in the development of abnormal behaviour. For example, confusing and unsupportive family environments, the societal inequalities that favour men over women, membership of a disadvantaged ethnic minority, or membership of a low socio-economic group may all contribute to the stress experienced by individuals who are biologically vulnerable to particular psychological conditions and who as a result develop psychological problems.

Second, much abnormal behaviour and experience is not neatly organized as a set of discrete syndromes with an under-lying biological cause. Boyle (1993) has shown, for example, that schizophrenia does not meet the specified criteria for a medical syndrome. Furthermore, much abnormal behaviour is normally distributed within the population. Thus dimensional rather than categorical conceptualizations of specific conditions are probably more valid for many conditions. For example, conduct disorder can probably be most validly conceptualized as instances where youngsters show extreme levels of externalizing behaviour problems, which fall on a continuum, with most children showing a moderate number of these types of difficulties (Achenbach, 1991). Also, trait conceptualizations of personality disorders are probably more valid than categorical conceptualizations.

Third, many psychopharmacological and physical treatments have harmful or unknown side-effects. For example, many widely used neuroleptic antipsychotic agents lead to an irreversible neurological movement disorder known as tardive dyskinesia (Kane, 1995). Electroconvulsive therapy (ECT), which is commonly used to treat major depression which is unresponsive to antidepressant medication, invariably leads to memory loss and confusion (Lock, 1999). The long-term effects of stimulant therapy such as Ritalin (methylphenidate), which is widely used to treat ADHD, are currently unknown (Cowart, 1988; Hinshaw, 1994).

Fourth, in some instances abnormal behaviour is a reaction to stresses inherent in family structures, such as oppressive, neglectful or abusive patterns of family organization. In others, abnormal behaviour is a reaction to stresses inherent in the structure of society at large, such as poverty, injustice, prejudice, racism, sexism, ageism and intolerance for non-conformity. The biological model, with its exclusive emphasis on organic factors in the etiology of psychological problems and its privileging of physical treatment, draws attention away from important psychosocial and political changes that may be required to alleviate psychological distress and to preserve civil liberties (Laing, 1961, 1965; Szasz, 1961, 1963; Newnes and

MacLachlan, 1996). By insisting, from the privileged position of a high-status profession, that abnormal behaviour is a reflection of mental illness requiring physical treatment, proponents of the biological model prevent society from addressing injustices, stresses and intolerance for non-conformity at a political level.

Fifth, the development of asylums has led to the exclusion of people with psychological problems from society. This in turn has promoted stigmatization and marginalization of large groups of people who engage in abnormal behaviour. The stigmatization and marginalization may further reinforce the difficulties of people with psychological problems and coping difficulties (Rosenhan, 1973; Newnes and MacLachlan, 1996).

Sixth, mental health legislation, which emerged within the biological and medical tradition for understanding abnormal behaviour, has been abused so as to limit the freedom of people with psychological problems. Szasz (1961, 1963) has argued that imprisoned criminals have more liberty and fewer violations of personal rights than people with psychological problems who are involuntarily detained. Seventh, because the biological model of abnormal behaviour and practices based upon it are championed by biologically oriented psychiatry, non-medical professions, including clinical psychology and psychotherapy, have often been marginalized in the development of mental health services.

Psychoanalytic Model

Assumptions

The psychoanalytic or psychodynamic model assumes that abnormal behaviour patterns are symptoms of underlying unconscious conflict or psychopathology (Wachtel and Messer, 1997; Tyrer and Steinberg, 1998). As a child develops, according to classical psychoanalytic theory, primitive sexual and aggressive urges of the unconscious 'id' become gradually controlled by the rational 'ego'. The ego is guided by an internalization of society's standards: the 'superego'. However, intrapsychic conflict is inevitable. Conflict occurs between the sexual and aggressive impulses of the id and societal standards as reflected in the superego.

Such conflict is managed unconsciously by using various defence mechanisms, the function of which is to keep the forbidden sexual and aggressive impulses from consciousness. For example, a man who is angry at his boss at work may sing his superior's praises, thereby using the defence of reaction formation. However, defences are compromises between the forces of the id and superego and often carry costly side-effects. For example, the man who is angry at his boss may eventually develop chest pains and anxiety, as a result of repressing rather than acknowledging the anger felt towards the superior.

Furthermore, the psychoanalytic model argues that relationship styles learned early in life are transferred in later life to other relationships, notably relationships with authority figures, sexual partners and psychotherapists. These relationship styles or 'transference phenomena' as they are called in psychoanalysis are coloured in part by feelings aroused and partially resolved during the Oedipus complex phase (for boys) or Electra complex phase (for girls). These developmental phases refer to the psychoanalytic hypothesis that children in early life desire their opposite-sex parent and harbour aggression towards the parent of their own gender. However, these sexual and aggressive impulses are repressed and the child eventually identifies with the parent of the same gender for fear of the consequences of acting them out. In later life, patients, such as the man with chest pains referred to earlier, experience feelings towards significant others and deal with them in a manner similar to that which occurred during the oedipal phase of development. So the man with chest pains experienced aggression towards his boss, and later towards his psychotherapist in a similar fashion, and dealt with this using reaction formation in a manner similar to that with which he handled the oedipal triangle as a child.

Psychoanalysis and psychoanalytic psychotherapy provide a context within which patients can experience transference towards a psychotherapist and then through interpretation gain insight into the transference and related defences which underlie their psychopathology. The analyst or therapist and client meet frequently according to a strict schedule. The patient reports in an uncensored way his or her contents of consciousness. Eventually the client develops transference and the analyst interprets this repeatedly over time until the patient has gained insight into the transference and related defences and worked through related unresolved feelings. Concurrently, the patient's symptoms abate. In order to be able to practise psychoanalysis, therapists must undergo their own analysis so that they have a first-hand understanding of the process, and so that they will recognize transference feelings that they have towards patients (counter-transference). Traditionally strict selection criteria are used for psychoanalysis and typically YAVIS (young, adult, verbal, intelligent and single) patients only have been deemed suitable.

ACHIEVEMENTS

The most outstanding achievement of the psychoanalytic model is the discovery of the unconscious (Ellenberger, 1970). Freud drew together a set of ideas from a wide range of sources and crystallized them in the notion of the unconscious, not as a passive repository of irretrievable memories but as an active set of psychological processes. According to the psychoanalytic model of the unconscious, people can make themselves forget things or keep them outside awareness.

Second, psychoanalysis gave meaning to apparently meaningless behaviour. For example, Freud showed how, in phobias, fears of one stimulus (e.g. one's father) could be displaced onto other stimuli (e.g. horses). He also showed how unconscious processes which explained psychological disorders could also explain peculiar everyday behaviour.

Third, psychoanalysis introduced the ideas of transference and countertransference into the practice of psychotherapy. This idea that people have a limited number of relationship-maps, which they learn early in life and transfer onto significant others in adulthood, has been supported by recent empirical work on continuity in patterns of attachment from childhood to adulthood (Feeney and Noller, 1996). Fourth, psychoanalysis established the place of the 'talking cure' in mainstream mental health practice. It also provided a model of the development of private long-term outpatient treatment of patients with moderately debilitating psychological difficulties. Finally, Freud provided a model for developing a theory of personality and therapy which spawned a wide range of neo-Freudian derivatives including those of Jung, Adler, Horney, Stack-Sullivan, Reich, Fairburn, Klein, Erikson and many others.

Limitations

Classical psychoanalysis has many limitations. First, many of its hypotheses were untestable due to the imprecision of the constructs or the imprecision of predictions entailed by the psychoanalytic theory. Also, for some considerable time there was little evidence for the effectiveness of psychoanalytic psychotherapy. In recent times, attempts have been made to address this by developing rating scales to evaluate defence mechanisms and formulations of core conflictual relationship themes, and by conducting controlled research trials to evaluate the effectiveness of psychodynamic psychotherapy (Miller *et al.*, 1993).

Second, Freud's speculations about infantile sexuality were not borne out by subsequent developmental research and in particular by research on child sexual abuse. It is quite likely that many of Freud's patients who reported sexual contact with a parent had in fact been sexually abused and were not simply fantasizing about seducing their parents in oedipal dramas (Masson, 1984).

Third, as a model for practice, classical psychoanalysis is too time-consuming and expensive. Classical psychoanalysis involves multiple sessions each week for a number of years. It is therefore not sufficiently cost-effective for routine use in a public mental health service where resources are limited. However, in recent times attempts have been made to use psychoanalytic ideas and practices as a basis for brief psychodynamic therapy (Messer and Warren, 1995; Holmes, 1999).

Cognitive-behavioural Model

Assumptions

The cognitive behavioural tradition incorporates a range of psycho-therapeutic theories and practices, including behaviour therapy, behaviour modification, cognitive therapy and cognitive-behaviour therapy, all of which have their roots in learning theories (Wachtel and Messer, 1997; Tyrer and Steinberg, 1998). Within the cognitive behavioural tradition, it is assumed that abnormal behaviour is a set of habits. Psychological disorders are clusters of learned behaviours, cognitions and affective states. Within the broad cognitive behavioural tradition it is assumed that abnormal behaviour is learned through the same processes as normal behaviour. These processes include operant and classical conditioning, imitation and insight.

Therapy involves coaching clients in such a way that they learn to replace abnormal with normal behaviour patterns. This coaching process is based on the principles of learning theory. So the small constituent habits that make up large clusters of behaviour, cognition and affect are identified through careful interviewing and observation. The antecedents, co-occurring beliefs, and consequences associated with individual habits are identified.

Treatment programmes include interventions that alter antecedents which signal the onset of abnormal behaviours; interventions that challenge non-adaptive beliefs and styles of information processing that accompany abnormal behaviour; and interventions that change the consequences of behaviour so that normal alternatives to abnormal behaviour patterns are reinforced. In the behavioural treatment of depression, antecedents of low mood may be altered through inviting clients to schedule pleasant events regularly throughout their day (Lewinsohn and Gotlib, 1995). In the cognitive therapy of depression, anxiety and personality disorders, clients are coached in how to identify and challenge negative automatic thoughts and to identify the use of cognitive distortions (Pretzer and Beck, 1996). With conduct-disordered children, reward programmes are used so that prosocial behavioural targets are routinely reinforced (Kazdin, 1995). Imitation and operant procedures may be used to help individuals with skills deficits learn social, communication and problem-solving skills. Skills training is routinely used when working with people who have schizophrenia and their families (Falloon *et al.*, 1993).

Treatment programmes also include procedures based on classical conditioning (Wachtel and Messer, 1997). For example, with systematic desensitization for people who have phobias, increasingly threatening stimuli are paired with the experience of relaxation. Another example of a classical conditioning-based intervention is the use of urine alarm programmes for nocturnal enuresis in which bedwetting is paired with the sound of an alarm. Through this procedure, the child learns eventually to awaken when the bladder is full.

Within the cognitive behavioural tradition, specific treatment programmes are developed for specific symptoms and detailed assessment of the impact of treatment on abnormal behaviours targeted in treatment is routinely made. This tradition is championed by clinical psychologists.

ACHIEVEMENTS

The cognitive behavioural tradition has made a number of important contributions to the understanding and treatment of psychological difficulties. First, this tradition has led to the development of a brief, effective approach to therapy which is applicable to a wide range of patients. Typically cognitive behaviour therapy is brief, ranging from 1 to 30 sessions, depending upon the nature and severity of the problems.

Second, the cognitive behavioural tradition has shown empirically that its treatment strategies are effective. It therefore facilitates evidence-based practice. More than any other approach to treating psychological difficulties, the cognitive behavioural tradition has generated an enormous volume of empirical research to test the effectiveness of a wide variety of treatment programmes for a broad range of problems in adults and children (Nathan and Gorman, 1998; Carr, 2000a).

Third, the cognitive behavioural model has led to the development of specific psychological treatment packages for specific types of problems. For example, exposure-based treatments have been developed for anxiety disorders, and cognitive therapy has been developed for mood disorders. Fourth, methodological and scientific rigour has characterized cognitive behavioural research on psychological problems and their treatment.

LIMITATIONS

The main limitation of the cognitive behavioural model is the risk it entails of trivializing psychological problems. When people are suffering and in profound psychological distress, it may seem to them and others that to construe their difficulties as bad habits does not do justice to the gravity of their situation.

Two other possible limitations deserve mention. First, there is the danger of disregarding the possible role of organic factors in the etiology of psychological difficulties, and second, there is the danger of not taking the role of the patient's wider social context into account. Having noted these two dangers, it should be mentioned that these have been addressed by various members of the cognitive-behavioural tradition at different times. For example, Eysenck (1979) in his conditioning model of neurosis argued that people with high levels of neuroticism and introversion were more conditionable and so more likely to develop phobias and that these two traits were 50 per cent heritable. Eysenck's position, therefore, while essentially behavioural, acknowledges the role of biological factors. With respect to social context, the work of Falloon *et al.* (1993) and others on reducing family stress in schizophrenia is a good example of social-context-sensitive cognitive behaviour therapy.

Family Systems Model

Assumptions

The family systems model assumes that psychological problems are maintained by patterns of interaction and belief systems within the family and the wider social system of the patient. Historical, contextual and constitutional factors may predispose family members to engage in these interaction patterns and adopt these belief systems. The many family therapy schools within this tradition may be classified in terms of their central focus of therapeutic concern and in particular with respect to their emphasis on (1) problem-maintaining behaviour patterns; (2) problematic and constraining belief systems; and (3) historical, contextual and constitutional predisposing factors.

With respect to the first theme, some family therapy schools highlight the role of repetitive patterns of family interaction in the maintenance of problem behaviour, and advocate practices which aim to disrupt these patterns of interaction. Schools that fall into this category include the MRI brief therapy approach (Segal, 1991); strategic therapy (Madanes, 1991); structural therapy (Colapinto, 1991); and functional family therapy (Barton and Alexander, 1981).

With respect to the second theme, some schools of family therapy point to the centrality of belief systems and narratives which subserve repetitive interaction patterns that maintain presenting problems. Practices that facilitate the emergence of new belief-systems and narratives which liberate family members from problem-maintaining interaction patterns are espoused by these schools. Schools that fall into this category include the Milan school (Campbell *et al.*, 1991); solution-focused family therapy (Wetchler, 1996); and narrative therapy (Freedman and Combs, 1996).

With respect to the third theme, a number of family therapy traditions highlight the role of historical, contextual and constitutional factors in predisposing family members to adopt particular belief systems and engage in particular problematic interaction patterns. Such schools advocate using practices that specifically address these historical, contextual and constitutional predisposing factors, including working with members of the extended family and wider social network as well as coaching individuals to manage historical, contextual and constitutional constraints. This category contains transgenerational family therapy (Friedman, 1991); psychoanalytic family therapy traditions (Scharff, 1995); attachment-theory-based approaches (Johnson and Greenberg, 1995); experiential family therapy (Wetchler and Piercy, 1996); multisystemic consultation, which includes reference to the wider system (Henggeler *et al.*, 1998), and psychoeducational approaches (McFarlane, 1991).

Elsewhere we have argued that an integrative approach to family therapy may be taken so that, for any problem, a formulation may be constructed using ideas from many schools of family therapy, in which the pattern of family interaction which maintains the problem is specified; the constraining beliefs and narratives that underpin each family member's role in this pattern are outlined; and the historical and contextual factors that underpin these belief systems and narratives are specified (Carr, 2000b). In parallel with this, a similar formulation may be constructed to explain why the problem does not occur in exceptional circumstances, which, while similar to problematic situations, differ in important key respects. In light of these formulations of the problem and exceptions to it, a range of interventions may be considered. Some interventions aim primarily to disrupt problem-maintaining behaviour patterns or amplify exceptional non-problematic patterns. Others aim to help family members change the personal narratives that make them repeat the same problematic behaviour patterns, and develop more liberating and flexible belief-systems which underpin exceptions to the problem. Still others aim to modify the negative impact of historical, contextual and constitutional factors or to draw on family strengths in these domains.

In family systems therapy, assessment and treatment involve the patient and the family participating in multiperson meetings. Multiple perspectives on the problem and related interaction patterns, belief systems and predisposing factors are therefore available to the therapist. Furthermore, there is the possibility of multiple people being involved in therapeutic change. Because of this, it is a fundamental assumption of family systems therapy that a small intervention may lead to a big change. It is therefore not surprising that family therapy is usually brief, with treatment rarely extending beyond twenty sessions. Furthermore, family therapy is dominated neither by clinical psychology nor psychiatry. Many disciplines are involved including social work and nursing.

ACHIEVEMENTS

Family therapy has made an important contribution to the understanding and treatment of abnormal behaviour. First, in a field dominated by essentially individualistic models of practice, it has highlighted the role of the patient's social context in the etiology and treatment of psychological difficulties. Second, family systems therapy is a brief, affordable form of treatment well suited to public health services. It is highly cost-effective. Where different family members have problems they may all be treated by the same therapist or team. Third, empirical research shows that family systems therapies are effective with a wide range of problems in children and adults (Carr, 2000b). Thus, there is a sound foundation for evidence-based practice. Fourth, systems theory can offer an integrative framework for comprehending not just the role of social factors but also those of biological and intrapsychic factors in the understanding and treatment of psychological difficulties. Fifth, in clinical practice, an integrative approach to family systems therapy is particularly useful in managing complex cases in which multiple family members have multiple problems, since often these are interconnected, a point missed by individualistic conceptualizations of abnormal behaviour.

LIMITATIONS

The main limitations of the family systems model are a danger of vagueness, a danger of losing sight of the individual, and a danger of failing to take account of organic factors. The biological, psychoanalytic, cognitive-behavioural and social systems models are each based on a unique set of assumptions, and have each resulted in a unique set of achievements; but each is limited in its usefulness by a specific set of shortcomings. With the biological model it is assumed that each set of abnormal behaviours constitutes the symptoms of an underlying brain disease, which follows a unique course, and has a particular prognosis, and for which a discrete cause and physical cure may ultimately be identified. The biological model's greatest achievement was the liberation of people with psychological problems and the creation of asylums where those in psychological distress received humane treatment. Mental health legislation, widely used classification systems such as the DSM and the ICD, a commitment to scientific study of abnormal behaviour, and the development of psychopharmacological treatments are among the major achievements of the biological model. The limitations of the biological model include the facts that much abnormal behaviour is not caused by organic factors; many psychopharmacological treatments have harmful side-effects; asylums have led to social exclusion; and mental health legislation has been abused.

The psychoanalytic model assumes that abnormal behaviour patterns are symptoms of underlying psychopathology. Psychotherapy gives insight into the conflicts, defences and transference phenomena which constitute this psychopathology. Achievements of the psychoanalytic model include the discovery of the unconscious; the demonstration of continuity between the normal and abnormal behaviour; and the establishment of talking therapy as a valid method for treating psychological problems. Classical psychoanalysis, however, is too expensive to be a viable approach for routinely treating people in the public health services. This and the fact that it is diagnostically vague, many of its propositions are untestable, and it pathologizes everyone are its major limitations.

The cognitive behavioural model assumes that symptoms are learned through conditioning, imitation and insight and that therapy involves changing patterns of learned behaviour using specific treatments that have been developed for specific symptoms. The development of a brief, effective approach to therapy, which is applicable to a wide range of patients, and for which there is empirical evidence for effectiveness, is the main achievement of the cognitive behavioural model. However, the risk of trivializing psychological problems and the risk of paying insufficient attention to organic or social factors are its principal limitations.

The systems model assumes that psychological problems are maintained by patterns of interaction and belief systems within the family and the wider social system of the patient. Historical, contextual and constitutional factors may predispose family members to engage in these interaction patterns and adopt these belief systems. In family systems therapy, assessment and treatment involve the patient and the family. In terms of achievements, empirical research shows that family systems therapies are effective with a wide range of problems and that systems theory offers a framework for integrating biological, psychological and social factors. However, the main limitations with this approach are a danger of vagueness, a danger of losing sight of the individual, and a danger of failing to take account of organic factors.

Glossary

Agoraphobia :

A fear of leaving the safety of the home which often occurs secondary to panic disorder.

Anomie :

A state characterized by normlessness where the social structures provided by family, religion and other institutions become destabilized and leave members of a community feeling alienated. Some disadvantaged communities in which delinquency, drug abuse and suicide occur are characterized by anomie.

Antisocial personality disorder :

A personality disorder characterized by a pervasive disregard for the rights of others and consistent violation of these rights. This pattern has also been referred to as psychopathy, sociopathy, and dissocial personality disorder.

Attention Deficit Hyper activity Disorder (ADHD) :

A syndrome characterized by persistent overactivity, impulsivity and difficulties in sustaining attention. Also known as attention deficit disorder, hyperkinetic disorder, hyperkinesis and minimal brain dysfunction.

Avoidant personality disorder :

A personality disorder characterized by a pervasive pattern of social inhibition and shyness beginning in adolescence or early adulthood.

Acquisition :

In Pavlovian conditioning, the learning of a response based on the contingency between a conditioned stimulus and an unconditioned stimulus.

Acute schizophrenia :

A condition characterized by the rapid and sudden onset of very florid symptoms of schizophrenia.

Acute stress disorder :

Anxiety, numbness, and disinterest, occurring during the first month after having experienced, witnessed, or been confronted by an event that involved the threat of death, injury, or threat to the physical integrity of self or others. If symptoms continue after the first month, it is called post-traumatic stress disorder.

Adoption study :

The study of offspring and biological and adoptive parents to see whether the occurrence of a trait is related to genes or environment. This method is one of two widely used methods (the other is the twin study) for quantifying the relative contribution of genetic to nongenetic determinants of psychological traits.

Affective contrast :

The subjective experience of positive and negative emotions based on drug exposure and withdrawal.

Affective pleasure :

An effect of drug use in the opponent-process model of addiction. Affective pleasure is the pleasant emotional state that is the initial pharmacological effect produced by the drug.

Affective tolerance :

An effect of drug use in the opponent-process model of addiction. With continued use, the addictive drug tends to lose its affective pleasure.

Affective withdrawal :

An effect of drug use in the opponent-process model of addiction. The sudden termination of narcotic use often produces the opposite affective state of the initial pleasant one.

Agonists :

Drugs that increase activity in particular neurotransmitter systems.

Agoraphobia :

An anxiety disorder characterized by fear of situations in which one might be trapped and unable to acquire help, especially in the event of a panic attack. People with agoraphobia will avoid crowds, enclosed spaces (such as elevators and buses), or large open spaces. From the Greek "fear of the marketplace."

Akathisia :

A side effect of chlorpromazine in which an itchiness in the muscles occurs and results in an inability to sit still.

Alarm reaction :

The first stage of a person's reaction to stress, according to Selye's theory of a general adaptation syndrome. The physiological arousal prepares the body to fight or flee the stressor.

Alexithymia :

The word literally means "no words for feelings" and is used to describe people who have difficulty expressing their feelings.

Allele :

One of the forms of a gene pair.

Allostasis :

The adaptation of many bodily states simultaneously across many different life circumstances.

Allostatic load :

The hidden price the body pays for its mediation of stress in the short run. High allostatic stress happens when there is frequent stress, repeated stress, or the inability to shut off stress mediators when they have been aroused.

Altruistic suicide :

Suicide required by the society (as defined by Durkheim; for example, hari-kari).

Alzheimer's disease :

Degenerative disease of late middle or old age, in which mental functions deteriorate. An amnesic syndrome is often the major feature of this disorder. Its initial symptoms include loss of initiative, extreme forgetfulness, memory disability, and spatial disorders. It progresses to severe deteriorations of intellect and basic maintenance functions that lead to death.

Amenorrhea :

Loss of the menstrual period. A common occurrence in women who have anorexia.

American Law Institute (ALI) rule :

A legal test for insanity that holds that a person is not responsible for his criminal conduct if, at the time of the crime and as a result of mental disease or defect, he lacks substantial capacity to appreciate the criminality of the conduct or to conform his conduct to the law.

Amnesia :

A dissociative disorder characterized by loss of memory of happenings during a certain time period, or loss of memory of personal identity. Includes generalized amnesia, retrograde amnesia, post-traumatic amnesia, anterograde amnesia, and selective amnesia.

Amphetamine :

A stimulant that causes agitation, increase in energy and activity, hyper-responsiveness to the environment, euphoria, and a number of physiological signs of hyperactivation.

Anaclitic depression :

A depression experienced by some infants between the ages of six and eighteen months who have been separated from their mothers for prolonged periods. This disorder is characterized by apathy, listlessness, weight loss, susceptibility to illness, and sometimes death.

Androgen :

A hormone that is principally responsible for the morphological development of the external genitals of the male.

Androgen insensitivity syndrome :

A syndrome in which a chromosomally male fetus lacks the gene that enables its body to recognize and react to masculinizing hormones, leading to an insensitivity to masculinizing hormones and the formation of female external genitals, rather than male external genitals.

Animal phobias :

Specific phobias in which an individual has a fear of a particular animal, usually cats, dogs, birds, rats, snakes, and insects.

Animism :

The belief in premodern societies that everyone and everything has a "soul" and that mental disturbance was due to animistic causes, e.g., an evil spirit had taken control of an individual and was controlling that individual's behavior.

Anomic suicide :

Suicide precipitated by a shattering break in an individual's relationship to his society (as defined by Durkheim).

Anorexia nervosa :

A disorder in which the individual has an intense fear of becoming fat, eats far too little to sustain herself, and has a distorted body image.

Antagonists :

Drugs that reduce the activity of a particular neurotransmitter by interfering with synthesis or blocking postsynaptic uptake of the neurotransmitter.

Anterograde amnesia :

Difficulty learning new material after a traumatic event.

Antigens :

Invaders of the immune system.

Antisocial personality disorder (or psychopathy, sociopathy) :

A personality disorder in which the individual has a rapacious attitude toward others and a chronic insensitivity and indifference to the rights of other people. The behavior must be long-standing and must be manifested in at least three classes of behavior, among which are: repeated aggressiveness; recklessness that endangers others; deceitfulness, lack of remorse, and consistent irresponsibility; failure to honor financial obligations.

Anxiety :

Fear characterized by the expectation of an unspecified danger, dread, terror, or apprehension, often leading to an emergency reaction and "flight or fight" behavior. As used in psychoanalytic theory, the psychic pain that results from conflicts among the various personality processes.

Anxiety disorders :

A class of mental disorders characterized by chronic and debilitating anxiety. These include agoraphobia, generalized anxiety disorder, panic disorder, phobias, post-traumatic stress disorder and obsessive-compulsive disorder.

Aphasia :

Disorders of language resulting from damage to certain areas of the cerebral cortex.

Applied tension :

A technique in which, upon seeing blood, people with a blood phobia tense the muscles of their arms, legs, and chest, thereby raising their blood pressure and heart rate, which prevents them from fainting at the sight of blood.

Appraisal :

Evaluation of short-term mental events, a target of cognitive therapy.

Apraxia :

A disorder of movement in the absence of muscle weakness or inability to perform any specific movement.

Arbitrary inference :

Reaching a conclusion for which there is little or no evidence. According to Beck, people who are depressed are prone to making arbitrary inferences.

Archetypes :

As used by Jung, universal ideas about which we are knowledgeable even at birth.

Asperger's disorder :

A pervasive development disorder that is usually not detected until the preschool period or later. It involves social impairment and repetitive patterns of behavior and limited interests.

Association areas :

Sectors of the brain that serve to integrate information from diverse sources.

Attribution :

An assignment of cause for an event; a short-term mental event, and a target of cognitive therapy.

Autism :

A childhood disorder whose central feature is the failure to develop the ability to respond to others within the first thirty months of life.

Automatic thoughts :

Discrete sentences, negative in character, that a person says to himself, quickly and habitually. According to Beck, people who are depressed typically engage in automatic thoughts.

Aversion therapy :

A behavior therapy that seeks to rid a client of undesired behavior by pairing that behavior with aversive consequences.

Avoidance-approach conflict :

A conflict between a desire to approach an object or situation that has some positive value, and a desire to avoid that object or situation because it has been associated with harm. According to traditional learning theory, this conflict is a root of anxiety.

Avoidance responding :

The act of getting out of a situation that has been previously associated with an aversive event, thereby preventing the aversive event. Differs from escape responding, which is getting out of the aversive event itself.

Avoidant personality disorder :

A disorder whose central feature is social withdrawal combined with hypersensitivity to rejection.

Babinski reflex :

A reflex reaction, normally shown only by infants, in which the big toe turns upward and the toes fan out when stroked.

Barbiturates :

A class of drugs depressing the central nervous system, decreasing anxiety and blunting sensitivity to the environment. Includes phenobarbital, pentobarbital, secobarbital, and benzodiapines.

B cells :

Lymphocytes that come from bone marrow and that have receptors on their surface for specific antigens.

Behavioral Activation System (BAS) :

A circuit connecting subcortical and cortical regions of the brain. It has been proposed that this system, which is sensitive to signs of pleasure and reward, inhibits the activity of the BIS.

Behavioral assessment :

A record of behaviors and thoughts one wishes to change, including their time of occurrence, duration, and intensity.

Behavioral disorder :

A disorder in which something behavioral, rather than emotional, is amiss, such as hyperactivity, attentional problems, and aggressive, destructive, and dishonest behaviors.

Behavioral Inhibition System (BIS) :

A circuit connecting subcortical and cortical regions of the brain. It is believed to play an important role in the experience of anxiety.

Behavioral model :

The school of abnormal psychology that claims that behavior is shaped by the environment, and that behavior can be changed by changing the environment. According to the behavioral theorists, the symptomatic behavior of a mental disorder is the disorder, and is that which should be treated.

Behavioral therapy :

A therapy that is rooted in the view that psychological distress results from learned behavior that can be unlearned; the therapy seeks to replace the distressing behavior with more constructive modes of coping and adaptation.

Behaviorism :

A movement in academic psychology from the 1920s to the mid-1960s that sought to discover in the laboratory the general laws of learning and to apply these laws to society as a whole.

Benzodiazepines :

A group of mild tranquilizers producing muscle relaxation, decreased anxiety, and sedation. Includes Librium, Valium, and Dalmane.

Biofeedback :

Therapeutic technique in which the individual is given electronically amplified information on certain (somewhat) controllable physiological systems (such as heart rate and blood pressure) and trained to control that response system.

Biogenic amines :

Neurochemicals that facilitate neural transmission, including catecholamines and indoleamines.

Biomedical model of abnormality :

The school of abnormal psychology that claims that mental disorders are illnesses of the body resulting from an underlying physiological pathology such as a virus, disordered biochemistry or genes, or a dysfunctional organ.

Bipolar disorder (or manic-depressive) :

An affective disorder characterized by alternating periods of depression and mania.

Bisexuality :

Desire for sexual relations with members of both sexes.

Blitz rape :

Sudden and unexpected rape, as contrasted to acquaintance rape, often leading to PTSD.

Blood-brain barrier :

A barrier in the brain composed of specialized cells that prevent certain compounds in the circulatory system from entering the brain.

Blood phobia :

A specific phobia in which the individual becomes highly anxious in situations involving the sight of blood, injections, and injuries.

Body Dysmorphic Disorder (BDD) :

The exaggeration of a slight bodily defect into the perception of wholesale ugliness, the preoccupation with which comes to dominate the person's life.

Borderline personality disorder :

A broad Axis-II diagnostic category that designates people whose salient characteristic is instability in a variety of personality areas, including interpersonal relationships, behavior, mood, and self-image.

Brain imaging :

Techniques such as PET scans and CAT scans that capture the way that the brain looks and functions.

Breakdown :

One of the ways in which norepinephrine is reduced during neural transmission; the enzyme monoamine oxidase (MAO) and other enzymes chemically break down the norepinephrine in the synaptic gap and render it inactive.

Blimia :

A disorder in which people alternately gorge themselves with enormous quantities of food, and then purge themselves of that food by vomiting, or using laxatives or diuretics.

Basal ganglia :

A subcortical region of the brain which subserves the initiation, control and modulation of voluntary movement. Damage to the basal ganglia may lead to excessive involuntary movement or to a slowing of voluntary movement. Parkinson's disease, Sydenham's chorea and Huntington's chorea all involve damage to the basal ganglia. Dysfunction of the basal ganglia underpins obsessive-compulsive disorder and Tourette syndrome.

Bipolar disorder :

A recurrent mood disorder involving episodes of depression and mania. The depressive episodes are characterized by low mood, negative cognition, and sleep and appetite disturbance, and the manic episodes are characterized by elation, grandiosity, flight of ideas and expansive behaviour.

Borderline personality disorder :

A personality disorder characterized by fear of abandonment, impulsivity and a pattern of pervasive instability in interpersonal relationships, self-image and mood.

Catatonic schizophrenia :

A form of schizophrenia in which there is alternation between extremes of excitability and stupor; automatic obedience or negativism; peculiar postures; and waxy flexibility of the limbs.

Comorbid :

Having more than one diagnosed psychological disorder or problem at a time, for example, schizophrenia and depression.

Conduct disorder :

A pervasive and persistent pattern of antisocial behaviour which extends beyond the family to the school and community and involves serious violations of rules, which is characterized by defiance of authority, aggression, destructiveness, deceitfulness and cruelty.

Conversion hysteria :

The symptoms, which are not feigned, include deficits in sensory or motor functioning in the absence of an organic illness. In ICD 10 and DSM IV, hysteria is now termed conversion disorder.

Cyclothymia :

A mood disorder like bipolar disorder with extreme mood fluctuations.

Caffeine :

A drug that stimulates the central nervous system and the skeletal muscles, lengthening the time it takes to fall asleep, decreasing fatigue, and aiding the individual in doing physical work.

Catatonic schizophrenia :

A subtype of schizophrenia that involves motor behavior that is either enormously excited or strikingly frozen.

Catecholamines :

Hormones involved in neural transmission in the brain. Includes norepinephrine, epinephrine, and dopamine.

Categorical approach :

An approach to diagnosis (and diagnostic manuals), in which the person either meets the behavioral criteria and falls in the category, or fails to meet the criteria and is deemed not to have that particular disorder.

Catharsis :

In psychoanalytic theory, the uncovering and reliving of early traumatic conflicts.

Cathexis :

The charging of a neutral object with psychic energy, either positive or negative.

Central Nervous System (CNS) :

That part of the nervous system that coordinates all of the activity of the nervous system. In vertebrates, the CNS is made up of the brain and the spinal cord. All sensory inputs are transmitted to the CNS and all motor impulses are transmitted from the CNS.

Cerebral cortex :

The outermost layer (gray matter) of the cerebral hemispheres.

Childhood disintegrative disorder :

A pervasive development disorder that resembles autism. Symptoms do not appear until the child reaches at least two years of age, at which time he or she begins to show a loss of skills in two of three areas: language, social, and motor.

Chromosomes :

The DNA-containing bodies that contain the individual's genes.

Chronic depression :

A depressive state in which the individual has both a chronic depression and an episodic depression on top of that.

Chronic schizophrenia :

A condition characterized by a prolonged and gradual period of decline in which no specific crisis or stressor can be identified.

Civil commitment :

The process used to hospitalize mentally disordered people who have committed no crime.

Clang associations :

Associations produced by the rhyme of words. Commonly found in people who have schizophrenia.

Clinical case history :

The record of part of the life of an individual seen in therapy.

Clinical interview :

The most common method of assessment. Clinicians use this approach to get information about a patient's psychological state by listening to what the patient says and how he or she says it.

Cocaine :

The psychoactive agent in the coca plant. Cocaine increases energy, combats fatigue and boredom, and enhances the individual's responsiveness to things in his environment.

Cognitions :

Beliefs, thoughts, attitudes, expectations, and other mental events.

Cognitive-behavioral therapy :

A therapeutic technique in which therapists attempt to alter both the maladaptive thoughts and maladaptive behaviors of a client through restructuring of maladaptive belief systems and re-training behavior.

Cognitive (or selective) filter :

A mechanism for sorting out stimuli to determine which ones will be admitted and which ones barred; something seems wrong with this filter in people who have schizophrenia.

Cognitive model :

The school of abnormal psychology that claims that many disorders result from maladaptive beliefs or thought styles.

Cognitive rehabilitation :

A treatment focussing on attention, memory, and executive functions.

Cognitive restructuring :

Treatments that are predicated on the assumption that irrational thoughts create irrational behaviors, which can be eliminated by changing the underlying thoughts.

Cognitive therapy :

Used primarily in the treatment of depression, this therapy seeks to change the cognitive triad of (a) self-devaluation, (b) a negative view of life experience, and (c) the pessimistic view of the future, as the determining cognitions for depression.

Cognitive triad :

According to Beck, a group of cognitions characterizing people diagnosed with depression. These cognitions include (a) negative thoughts about the self, (b) negative thoughts about ongoing experience, and (c) negative thoughts about the future.

Collective unconscious :

As used by Jung, the memory traces of the experience of past generations.

Comorbidity :

The co-occurrence of two or more diagnoses.

Comparative studies :

A type of study that contrasts two or more groups in order to find out how people with a particular symptom or disorder differ from people without the symptom or disorder.

Compulsion :

A repetitive, stereotyped, and unwanted action that can be resisted only with difficulty. It is usually associated with obsessions.

Computerized-axial tomography (CAT or CT) scan :

An X-ray technique, used in neurological diagnosis, for constructing three-dimensional representations of the X-ray density of different areas of the brain.

Concordant :

When both of two twins have a disorder such as schizophrenia, they are called concordant for that disorder.

Conditioned Response (CR) :

A response that is evoked by a certain stimulus (conditioned stimulus) once that stimulus has become associated with some other stimulus (unconditioned stimulus) that naturally evokes the unconditioned response.

Conditioned Stimulus (CS) :

A stimulus that, because of its having been paired with another stimulus (unconditioned stimulus) that naturally provokes an unconditioned response, is eventually able to evoke that response.

Conduct disorders :

A cluster of children's behavioral disorders consisting mainly of aggressive and rule-breaking behaviors.

Confound :

A factor other than the experimentally controlled independent variable that **might produce an** experimental effect.

Congenital Adrenal Hyperplasia (CAH):

A disease that bathes female fetuses with extra androgen. As young children, CAH girls like male-stereotyped toys, rough and tumble play, and are more tomboyish than matched controls.

Construct validity :

The extent to which diagnostic tests facilitate communication by describing and categorizing patients.

Contingency :

A conditional relationship between two objects or events, describable by the probability of event A given event B, along with the probability of event A in the absence of event B. A positive contingency between A and B obtains when A is more likely in the presence of B than in the absence of B.

Control group: :

A group of subjects similar to those in an experimental group, who experience everything the experimental group does, except the independent variable.

Conversion :

A somatoform disorder characterized by the loss of functioning of some part of the body not due to any physical disorder, but apparently due to psychological conflicts. The loss is not under voluntary control.

Coping strategies :

As used by psychoanalytic theorists, the process by which people alter the meaning and significance of troublesome drives and impulses in order to eliminate anxiety.

Coprolalia :

The repetition of socially unacceptable words sometimes manifested by people with Tourette's disorder.

Core self :

The self that develops first, between the second and sixth month of an infant's life. It embraces the infant's awareness that she and her caregiver are physically separate.

Coronary Heart Disease (CHD) :

Heart or circulatory problems often caused by arteriosclerosis; it can often lead to heart attack and sudden death.

Correlation :

Pure observation without manipulation to determine the relationship between two classes of events.

Correlational studies :

A type of study in which experimenters observe the relationship between two or more factors as they exist in the real world.

Correlation coefficient :

A statistic indicating the degree of contingency between two variables.

Co-twin :

As used in psychological research, one of a pair of twins whose sibling is seen at a psychiatric clinic in order to diagnose a psychological problem.

Counterconditioning :

A therapeutic technique for phobias in which a patient with a phobia is helped to relax while imagining fear-provoking situations (usually the least fear-provoking situation first, then gradually more and more fear-provoking situations). The relaxation response to the imagined situation is incompatible with the fear the patient has previously associated with the situation, and the fear is thus extinguished.

Covert sensitization :

A behavioral therapy for changing sexual interest; it uses imaged sexual stimuli followed by aversive US's to treat such paraphilias as exhibitionism.

Criminal commitment :

The coerced psychiatric hospitalization of mentally disordered people who have acted harmfully but are not legally responsible because they lack *mens rea,* or a "guilty mind."

Cross tolerance :

When tolerance to one drug produces tolerance to other drugs.

Cultural-familial mental retardation :

A condition in which there is no identified biological basis for the retardation. Many researchers believe that this form of retardation is caused by insufficient intellectual stimulation from the child's environment.

Cyclothymic disorder :

A mild but chronic form of bipolar disorder in which the patient experiences depressive and hypomanic symptoms for at least two years.

Deep strategies of psychotherapy :

One of the "nonspecifics of therapy" used by all good therapists. They include such techniques as instilling hope and building buffering strengths in patients.

Degradation :

A mechanism that clears the synapse by eliminating excess neurotransmitter through the action of enzymes.

Deinstitutionalization :

A mass movement that released patients from psychiatric hospitals into the community at large.

Delayed auditory feedback :

A method of treatment for stuttering that involves hearing one's own speech played back at about a .1 second delay.

Delirium :

A cognitive disturbance that has a rapid onset, fluctuates over time, and usually responds to treatment.

Delirium tremens :

A dangerous syndrome of withdrawal from alcohol which is characterized by psychomotor agitation, hyperactivity of the autonomic nervous system, anxiety, loss of appetite, delusions, amnesia, and convulsions.

Delusional disorder :

A psychotic disorder in which the person experiences nonbizarre delusions for at least one month in duration.

Delusions :

False beliefs that resist all argument and are sustained in the face of evidence that normally would be sufficient to destroy them.

Delusions of control :

Beliefs that one's thoughts or behaviors are being controlled from without.

Delusions of grandeur :

Unsubstantiable convictions that one is especially important.

Delusions of persecution :

Groundless fears that individuals, groups, or the government has malevolent intentions and is "out to get me."

Delusions of reference :

Incorrect beliefs that the casual remarks or behaviors of others apply to oneself.

Demand characteristics :

Aspects of the experimental setting that induce the subject to invent and act on a hypothesis about how one should behave.

Dementia :

A more or less general deterioration of mental function, found most commonly in old people. Alzheimer's disease is a common form of dementia.

Denial :

As used in psychoanalytic theory, the process by which distressing external facts are eliminated.

Dependence syndrome :

A cluster of physiological, behavioral, and cognitive phenomena in which the use of a substance or a class of substances takes on a much higher priority for a given individual than other behaviors that once had higher value.

Dependent personality disorder :

A disorder wherein people allow others to make major decisions, to initiate important actions, and to assume responsibility for the significant areas of their life.

Dependent variable :

The factor that the experimenter expects will be affected by changes in the independent variable.

Depersonalization :

Feeling detachment from oneself; just going through the motions or looking at oneself from the outside.

Depression :

A mood disorder characterized by (a) sad affect and loss of interest in usually satisfying activities, (b) a negative view of the self and hopelessness, (c) passivity, indecisiveness, and suicidal intentions, and (d) loss of appetite, weight loss, sleep disturbances, and other physical symptoms.

Derealization :

Feeling as if the world, not the self, seems unreal.

Detoxification :

The reduction and removal of alcohol from the body.

Developmental disorders :

A cluster of disorders of childhood that may consist of deficits in language comprehension, speech, and responses to others that can result in such serious disorders as autism.

Diathesis :

Physical vulnerability or predisposition to a particular disorder.

Diathesis-stress model :

A general model of disorders that postulates that an individual develops a disorder when he both has some constitutional vulnerability (diathesis) and when he experiences psychological disturbance (stress).

Diffusion :

A mechanism that clears the synapse by reducing concentrations of neurotransmitter through an intermingling with other substances outside the neuron.

Dimensional approach :

An approach to diagnosis (and diagnostic manuals), in which the behaviors and traits that define personality disorders are believed to fall on a continuum.

Direct sexual therapy :

A therapeutic method developed by Masters and Johnson, in which (a) sexual dysfunctions are clearly and simply defined, (b) clients explicitly practice sexual behavior under the systematic guidance of therapists, and (c) clients are treated as couples, not as individuals.

Discordant :

When only one of two twins has a disorder such as schizophrenia, they are called discordant for that disorder.

Discriminative stimulus :

A signal that means reinforcement is available if the operant is made.

Disinhibition :

An increase in some reaction resulting from release of inhibition.

Disorganized schizophrenia :

A schizophrenic disorder whose most striking behavioral characteristic is apparent silliness and incoherence. Behavior is jovial but quite bizarre and absurd, suggesting extreme sensitivity to internal cues and extreme insensitivity to external ones, but without systematic delusions or hallucinations.

Displacement :

A cognitive alteration of reality that involves replacing the true object of one's emotions with one that is more innocent and less threatening.

Disruptive behavior disorders :

A cluster of disorders characterized by symptoms such as hyperactivity, inattention, aggressiveness, destructiveness, and defiance of authority.

Dissociation :

A situation in which two or more mental processes co-exist or alternate without being connected or influencing each other, with some area of memory split off or dissociated from conscious awareness.

Dissociative amnesia :

A loss of personal memory caused by severe trauma such as the death of a child or the dashing of a career. (Formerly known as psychogenic amnesia.)

Dissociative disorders :

A group of mental disorders characterized by fragmentation of an individual's identity. Dissociative disorders include amnesia, fugue, dissociative identity disorder, and depersonalization disorder.

Dissociative Identity Disorder (DID)
(or multiple personality disorder) :

The occurrence of two or more identities in the same individual, each of which is sufficiently integrated to have a relatively stable life of its own and recurrently to take full control of the person's behavior.

Dizygotic twins :

Fraternal twins, or twins who developed from separate eggs, and whose genes are no more alike than are those of any pair of non-identical-twin siblings.

DNA (Deoxyribonucleic Acid) :

The double helix molecule that contains the genes and serves as the blueprint for the synthesis of RNA, and the production of proteins and amino acids. The entwined strands of DNA are made up of phosphates, sugars, and four nucleotide bases.

Dominant gene :

Genes that can, on their own, produce a specific trait.

Dopamine :

A catecholamine that facilitates neural transmission.

Dopamine hypothesis :

The theory that schizophrenia results from an excess of the neurotransmitter dopamine.

Double-blind experiment :

An experiment in which both the subject and experimenter do not know whether the subject has received an experimental treatment or a placebo.

Double depression :

A major depressive episode superimposed on an underlying chronic depression.

Down syndrome :

A disorder that results from the fact that an individual has forty-seven rather than the usual forty-six chromosomes in his or her cells.

Drug cues :

Drug reminders, such as settings in which drugs were taken, that induce craving for the drug and can trigger relapse in those who have been free of drugs.

Drug dependence :

The regular use of drugs acting on the brain that leads to maladaptive behavioral changes that would be seen as maladaptive in any culture. Three criteria characterize the disorder: (a) a pattern of pathological use of a drug, (b) impairment in occupational, social, physical, or emotional functioning, and (c) evidence of affective or physical adaptation to the drug.

Drug tolerance :

The need to use increased amounts of the drug to get the desired effect.

Drug withdrawal :

Characteristic affective and physical symptoms that follow drug use after the drug is discontinued.

DSM-IV :

Published in 1994, this is the fourth edition of the *Diagnostic and Statistical Manual of Mental Disorders* of the American Psychiatric Association.

Durham test :

A legal test for insanity which provides that an accused is not criminally responsible if his unlawful act was the product of mental disease or mental defect.

Dysfunction :

Impairment of functioning.

Dyslexia :

A learning disorder that leads to a difficulty in learning to read.

Dysphoria :

An unpleasant emotional state experienced during drug withdrawal; the opposite of euphoria.

Dysthymic disorder :

Chronic depression in which the individual has been depressed for at least two years without having had a remission to normality that lasted more than two months.

Delusions :

Unfounded and culturally alien beliefs.

Dependent personality disorder :

A personality disorder characterized by a pervasive pattern of submissiveness and clinginess.

Neurotransmitters :

Chemicals released into synapses between neurons which permit messages to be transmitted from one neuron to the next. Dysregulations of neurotransmitters may occur in certain conditions. For example, there is dysregulation of the dopamine system in schizophrenia.

Obsessive-compulsive Disorder (OCD) :

A condition characterized by distressing obsessional thoughts or impulses on the one hand and compulsive rituals which reduce the anxiety associated with the obsessions on the other.

Obsessive-compulsive personality disorder :

A personality disorder characterized by a pervasive pattern of preoccupation with orderliness, perfectionism, ethics, interpersonal control, and fiscal economy. Referred to as anakastic personality disorder in ICD 10.

Oppositional defiant disorder :

A disorder of conduct confined to the home and characterized by difficulties with rule-following.

Panic disorder :

Recurrent unexpected panic attacks, that is, acute episodes of intense anxiety in which the person may experience a fear of losing control; a fear of going crazy; or a fear of dying. Commonly secondary agoraphobia develops.

Paranoid personality disorder :

A personality disorder characterized by a pervasive distrust of others.

Paranoid schizophrenia :

A form of schizophrenia in which paranoid delusions are the predominant feature.

Personality disorder :

Conditions characterized by enduring patterns of behaviour and experience that deviate markedly from cultural expectations and which lead to significant personal distress or significant impairment in social functioning.

Phobic anxiety :

Intense fear which occurs when faced with an object, event or situation from a clearly definable class of stimuli, which is out of proportion to the danger posed by the stimulus.

Post Traumatic Stress Disorder (PTSD) :

Following a catastrophic trauma which individuals perceived to be potentially life-threatening for themselves or others, a pattern characterized by recurrent intrusive memories of the trauma, which leads to intense anxiety coupled with attempts to avoid this by suppressing the memories and avoiding situations that remind them of the trauma.

Prodromal phase :

The period before the onset of a disorder during which individuals may show subclinical symptoms.

Schizoid personality disorder :

A personality disorder characterized by a pervasive pattern of detachment from social relationships.

Schizophrenia :

A set of seriously debilitating conditions characterized chiefly by hallucinations, delusions and thought disorder.

Schizotypal personality disorder :

A personality disorder characterized by unusual perceptual experiences, eccentric thoughts and speech, inappropriate or constricted affect, peculiar or eccentric behaviour, and a lack of close relationships.

Separation anxiety :

Inappropriate fear aroused by separation from an attachment figure.

SSRI :

Selective serotonin reuptake inhibitors (e.g. fluoxetine/Prozac), a type of antidepressant that prevents serotonin from being reabsorbed into the presynaptic membrane and so increases levels of this neurotransmitter.

TCA :

drug abuse and suicide occur are characterized by anomie. Tricyclic antidepressants, like imipramine/ Tofranil, which increase the sensitivity of dysfunctional receptor sites to neurotransmitters, particularly noradrenalin.

Thought disorder :

Confused thinking common in schizophrenia, characterized by disorganized and illogical speech.

Token economy :

A therapeutic system used in residential or inpatient settings, based on learning theory, in which tokens are used as secondary reinforcers to encourage patients to engage in positive behaviours and to use adaptive skills. Tokens earned for completing positive behaviours or using adaptive skills may be exchanged for privileges or valued items.

Validity :

The extent to which a test of something is actually measuring that something.

Vascular dementia :

The leading cause of irreversible dementia, it is often caused by a stroke and the resulting loss of oxygen to the brain kills brain tissue. The cognitive signs parallel those of Alzheimer's disease, but the patient often shows unusual reflexes and movement abnormalities.

Ventricles :

Cavities in the brain, containing cerebral spinal fluid.

Verbal self :

The third sense of self, which develops between fifteen and eighteen months of age. It is the verbal and symbolic storehouse of experience and knowledge.

Voyeurism :

A psychosexual disorder in which the individual habitually becomes sexually aroused only by observing the naked body, the disrobing, or the sexual activity of an unsuspecting victim.

Withdrawal syndrome (or abstinence syndrome) :

A substance-specific syndrome that follows cessation of the intake of a substance that has been regularly used by the individual to induce intoxication; usually characterized by observable, physical signs such as marked changes in body temperature or heart rate, seizures, tremors, or vomiting.

Yoking :

An experimental procedure in which both experimental and control groups receive exactly the same physical events, but only the experimental group influences these events by its responding.

Zygote :

A cell formed from the union of the egg and the sperm.

Bibliography

A. A. Brill, Psychanalysis, W. B. Saunders Co., 1914, p. 48. The case as here cited is much abbreviated and condensed.

A. Adler, Practice and Theory of Individual Psychology, Harcourt, Brace & Co., 1924, p.

A. Adler, The Neurotic Constitution, Moffatt, Yard & Co., 1917.

A. D. Tendler, "The Mental Status of Neurotics", *Archives of Psychology*, Columbia University Press, No. 60, 1923.

A. F. Hurst, The Psychology of the Special Senses and Their Functional Disorders, Croonian Lectures, 1920.

A. F. Riggs, "The Psychoneuroses: Their Nature and Treatment", *American Journal of Psychiatry*, 1923, pp. 94, 95.

A. J. Rosanoff et al., Manual of Psychiatry, John Wiley & Sons, (6th ed.) 1927, pp. 122, 142.

A. J. Rosanoff, "A Theory of Personality Types, Based on Psychiatric Experience", Psychological Bulletin, 1920.

A. J. Rosanoff, A Study of Hysteria, etc., Archives of Neurology and Psychiatry, October, 1919, II, pp. 419-460.

A. J. Rosanoff, *Manual of Psychiatry*, John Wiley & Sons, (6th ed.) 1927. Other contributors, deFursac, Hollingworth, Jarrett, Neyman, and Williams.

Abbreviated from E. J. Kempf, Psychopathology, C. V. Mosby Medical Book & Publishing Co., 1920, p. 436.

Adapted from a description by Kraepelin, in Clinical Lectures in Psychiatry, Wm. Wood & Co., 1906.

Adapted from A. Meyer and G. H. Kirby, *Notes of Clinics in Psychopathology*, p. 89.

Adapted from A. R. Diefendorf, *Clinical Psychiatry*, The Macmillan Co., 1915, which is "abstracted and adapted from the 7th edition of Kraepelin, *Lehrbuch der Psychiatrie*.

Adapted from E. Kraepelin, Lectures in Clinical Psychiatry, Wm. Wood & Co., 1906.

Adapted from E. Kraepelin, *Lectures on Clinical Psychiatry*, Wm. Wood & Co., (2nd ed.) 1906, pp. 231-234.

Adapted from L. S. Hollingworth, The Psychology of the Adolescent, D. Appleton & Co., 1928, Ch. III on Psychological Weaning.

B. M. Morrison, A Study of the Major Emotions in Persons of Defective Intelligence, University of California Publications in Psychology, III, 3, December, 1914.

B. Sidis, *Causation and Treatment of Psychopathic Diseases*, R. G. Badger, 1916, III-IV.

B. T. Baldwin, "Distribution of School Training of Wounded Soldiers", *School and Society*, X, 258, p. 680.

Barlow, David H.; Vincent Mark Durand (2004). *Abnormal Psychology: An Integrative Approach*. Thomson Wadsworth

Bennett, Paul (2003), *Abnormal and Clinical Psychology*, Open University Press

Binet and Simon, The Intelligence of the Feeble-Minded, Dept. of Research, Training School at Vineland, N. J., 1916.

Binet and Simon, The Intelligence of the Feeble-Minded, Dept. of Research, Training School at Vineland, N. J., 1916, p. 112.

British Psychiatric Morbidity Survey, Jenkins et al. (1998), *The British Journal of Psychiatry* 173: 4-7.

C. Burt, Mental and Scholastic Tests, London, 1921, p. 335.

C. Burt, Mental and Scholastic Tests, London, 1921.

C. G. Jung, Psychological Types, Harcourt, Brace & Co., 1923.

Carver, Journal of Comparative Psychology, August, 1922.

Certain of the descriptions and instances in the following pages are quoted from the chapter on Abnormal Development in the writer's book, Mental Growth and Decline, D. Appleton & Co., 1927.

David H. Barlow and Vincent Mark Durand (2004). Abnormal Psychology: An Integrative Approach. p. 7

David H. Barlow and Vincent Mark Durand (2004). Abnormal Psychology: An Integrative Approach. p. 8

David H. Barlow and Vincent Mark Durand (2004). Abnormal Psychology: An Integrative Approach. p. 11

David H. Barlow and Vincent Mark Durand (2004). Abnormal Psychology: An Integrative Approach. p. 26

Dejerine and Gauckler, *Psychoneuroses and Their Treatment by Psychotherapy*, (Trans. by Jelliffe) J. B. Lippincott Co., 1915, p. 4. The italics are mine.

Diathesis stress model for panic-related distress: a test in a Russian epidemiological sample

E. A. Doll, Anthropometry as an Aid to Mental Diagnosis, Training School, Vineland, N. J., No. 8, February, 1916.

E. D. McDowell, *Educational and Emotional Adjustments of Stuttering Children*, Teachers College, N. Y. No. 314, 1928.

E. Evans, *The Nervous Child*, Dodd, Mead & Co., 1920.

E. Kretschmer, Physique and Character, Harcourt, Brace & Co., 1921.

E. T. Sullivan, A Comparison of Accelerated, Average and Retarded Fourth Grade Children, Journal of Delinquency, September-November, 1923.

F. Bambi Salten, *A Life in the Woods*, Simon and Schuster, 1928, p. 260.

F. H. Bradley, Principles of Logic, G. E. Stechert & Co., 1905. Bk. II, Pt. 2, Ch. 1.

F. L. Wells, "Mental Regression, Its Conception and Types", Psychiatric Bulletin, October, 1916.

F. L. Wells, Mental Adjustments, D. Appleton & Co., 1916, p. 151.

F. L. Wells, Mental Adjustments, D. Appleton & Co., 1916; also, "Von Bechterev und Ueberträgung", Journal of Philosophy, Psychology and Scientific Methods, June, 1916.

F. L. Wells, Mental Adjustments, D. Appleton & Co., 1917, pp. 137-138.

F. W. Mott, *War Neuroses and Shell Shock*, Oxford University Press, 1919.

Farrar, American Journal of Insanity, April, 1917.

For details of the method see H. L. Hollingworth, "Correlations of Achievement Within an Individual", Journal of Experimental Psychology, VIII, 3, June, 1925.

From Meyer and Kirby, Notes of Clinics in Psychopathology, pp. 67-68.

G. F. Herbart, *Text Book of Psychology*, D. Appleton & Co., 1816

G. H. Makuen, "A Brief History of the Treatment of Stammering", *Philadelphia Medical Journal*, Vol. XIII, 1909-1910, pp. 191197.

G. V. Hamilton, Objective Psychopathology, C. V. Mosby Medical Book & Publishing Co., 1925, p. 296.

G. V. Hamilton, Objective Psychopathology, C. V. Mosby Medical Book & Publishing Co., 1925, p. 297.

H. Aphasia Head, *Studies in Neurology*, Vol. I, Oxford University Press, 1920, p. 97.

H. C. Bastian, *Aphasia and Other Speech Defects*, London, 1808, is the most useful reference for this author.

H. C. Cotton, *The Defective, Delinquent and Insane*, Princeton University Press; "The Etiology and Treatment of the SoCalled Functional Neuroses", *American Journal of Psychiatry*, October, 1922.

H. Ellis, Man and Woman, Chas. Scribner's Sons, 1909 ed.

H. H. Woodrow, Journal of Experimental Psychology, August, 1916.

H. Head, Aphasia and Kindred Disorders of Speech, Cambridge University Press, 1926, Vol. I, p. 119.

H. Head, Aphasia and Kindred Disorders of Speech, Cambridge University Press, 1926.

H. L. Hollingworth, "The Influence of Caffein on Efficiency", Archives of Psychology, No. 22, 1912. "The Influence of Alcohol", Journal of Abnormal and Social Psychology, XVIII, 3 and 4, October, 1923 and January, 1924.

H. L. Hollingworth, "Variations in Efficiency During the Working Day", Psychological Review, November, 1914.

H. L. Hollingworth, Mental Growth and Decline, D. Appleton & Co., 1927, p. 176.

H. L. Hollingworth, Mental Growth and Decline, D. Appleton & Co., 1927, p. 59.

H. L. Hollingworth, Psychology of Functional Neuroses, D. Appleton & Co., 1920, pp. 50-51.

H. L. Hollingworth, Psychology of Functional Neuroses, D. Appleton & Co., 1920, pp.

H. L. Hollingworth, Psychology of Functional Neuroses, D. Appleton & Co., 1920, Ch. 3. Rosanoffet al., Manual of Psychiatry, John Wiley & Sons, (6th ed.) 1927, p. 165. (See note 5.)

H. L. Hollingworth, Psychology of Functional Neuroses, D. Appleton & Co., 1920, Ch. 6.

H. L. Hollingworth, Psychology of Functional Neuroses, D. Appleton & Co., 1920, p. 20.

H. L. Hollingworth, *Psychology of Functional Neuroses,* D. Appleton & Co., 1920.

H. L. Hollingworth, *Psychology of Functional Neuroses,* D. Appleton & Co., 1920, p. 89.

H. L. Hollingworth, *Psychology, Its Facts and Principles,* D. Appleton & Co., 1928.

H. Stelzner, Die Psychopathischen Konstitutionen, Karger, Berlin, 1912.

Hansell, James; Lisa Damour (2005). *Abnormal Psychology.* Von Hoffman Press

Hockenbury, Don; Sandra Hockenbury (2003). *Psychology* (3rd ed.). Worth Publishing. 129-1

Huxley Lecture, British Medical Journal, October, 1906.

I. Pavlov, Conditioned Reflexes, Oxford University Press, 1927, p. 27.

I. Pavlov, The Work of the Digestive Glands, J. B. Lippincott Co., 1911. Also Conditioned Reflexes, Oxford University Press.

J. B. Watson, Behavior and the Concept of Mental Disease, Journal of Philosophy, Psychology and Scientific Methods, October 26, 1916.

J. B. Watson, Psychology from the Standpoint of a Behaviorist, J. B. Lippincott Co., 1929, p. 419.

J. Hinshelwood, Congenital Word Blindness, Lewis & Co., London, 1917.

J. M. Fletcher, *The Problem of Stuttering*, Longmans, Green & Co., 1928.

J. Rosett, Archives of Neurology and Psychiatry, June, 1923.

J. V. Haberman, Medical Record, December 1, 1917.

J. W. Bridges, Outline of Abnormal Psychology, R. G. Adams & Co., 1921, p. 127.

James Hansell and Lisa Damour. Abnormal Psychology Ch 3. pg 37.

James Hansell and Lisa Damour. Abnormal Psychology. Ch 3. pg 30-33.

James Hansell and Lisa Damour. Abnormal Psychology. Ch 3. pg 37.

K. Dunlap, *Journal of Abnormal Psychology*, April-May, 1917.

K. S. Lashley, *Brain Mechanisms and Intelligence*, University of Chicago Press, 1929.

Kopeloff and Cheney, "Studies in Focal Infection, Its Presence and Elimination in the Functional Psychoses", *American Journal of Psychiatry*, October, 1922.

L. A. Tirapegui, Comparative Variability in Anthropometric Traits of Normal and Feeble-Minded, Lebreria e Imprenta, Santiago, Chile.

L. S. Hollingworth, and M. V. Cobb, Children Clustering at 165 I.Q. and Children Clustering at 146 I.Q. Compared for Three Years in Achievement, National Society for the Study of Education, 27th Year Book, 1928, Part II, pp. 3-33.

L. S. Hollingworth, *The Psychology of Subnormal Children*, The Macmillan Co., 1920.

Leta S. Hollingworth, "Differential Action Upon the Sexes of Forces which Tend to Segregate the Feeble Minded", Journal of Abnormal and Social Psychology, April-June, 1922.

Leta S. Hollingworth, Gifted Children, The Macmillan Co., 1926, and various other studies by the same author.

Lucretius, De Rerum Natura, Mason Good's translation, The Macmillan Co

M. L. Goodhue, *The Cure of Stage Fright*, Four Seas Co., 1927, p. 20.

M. Meyer, *Abnormal Psychology*, Lucas Bros., 1927.

M. Prince, Association Neuroses, etc., *Journal of Nervous and Mental Diseases*, May, 1891, p. 275.

M. Trumper, *A Hemato-Respiratory Study of 101 Consecutive Cases of Stammering*, University of Pennsylvania, 1928.

N. Norsworthy, "The Psychology of Mentally Deficient Children", *Archives of Psychology*, No. 1, November, 1906, p. 1.

N. Norsworthy, Psychology of Mentally Deficient Children, Science Press, 1906.

P. Janet, Major Symptoms of Hysteria, The Macmillan Co., 1907, p. 15.

P. Janet, The Mental State of Hystericals, G. P. Putnam's Sons, 1907.

P. Janet, The Mental States of Hystericals, G. P. Putnam's Sons, 1907, pp. 527-528.

Prince M., Association Neuroses, etc., Journal of Nervous and Mental Diseases, May, 1891, Vol. XVI, pp. 266-282.

Prince M., The Unconscious, The Macmillan Co., (2nd ed.) 1921.

Prince M., *The Unconscious*, The Macmillan Co., 1921, p. 368.

R. G. Rows, "The Biological Significance of Mental Illness", an address published in "A Psychiatric Milestone" in connection with the Bloomingdale Centenary, 1921, pp. 93-113.

R. Gaupp, Preface to first edition (Trans.) of Kretschmer, Physique and Character, Harcourt, Brace & Co., 1921.

R. S. Woodworth, Dynamic Psychology, Columbia University Press, 1918, pp. 157-161.

R. S. Woodworth, Psychology, Henry Holt & Co., (1st ed.) 1908, p. 553.

Readings in Abnormal Psychology and Mental Hygiene, edited by W. S. Taylor, D. Appleton & Co., 1926.

Reprinted from E. A. Doll, vide supra, p. 66.

Rosanoff*et al.*, *Manual of Psychiatry*, p. 28

S. Freud, Totem and Taboo, Moffatt, Yard & Co., 1918, pp. 45-49.

S. Freud, Totem and Taboo, Moffatt, Yard & Co., 1918, pp. 45-49.

Sentence quoted from A. J. Rosanoff, *Manual of Psychiatry*, John Wiley & Sons, (6th ed.) 1927.

The case and its explanation are quoted from a paper by Dr. Ernest Jones, Journal of Abnormal Psychology, June, 1913.

The Lancet, London, June 9, 1917.

This case is abstracted and abbreviated from I. Coriat, Abnormal Psychology, Moffatt, Yard & Co., 1914, p. 229.

Thus stated by A. A. Brill, Psychanalysis, W. B. Saunders Co., 1914.

Understanding the multiple causality of psychopathology

W. A. White, *Outlines of Psychiatry*, Journal of Nervous and Mental Disease Publishing Co., (5th ed.) 1915, Ch. 3.

W. A. White, Outlines of Psychiatry, Nervous and Mental Disease Monograph Series, No. 1, (5th ed.) 1915, p. 99.

W. B. Cannon, *Bodily Changes in Pain, Hunger, Fear and Rage*, D. Appleton & Co., 1915.

W. Dennis, Observations on Two Responses of the White Rat, Journal of Genetic Psychology, December, 1928.

W. H. R. Rivers, "War Neurosis and Military Training", *Mental Hygiene*, October, 1918.

W. H. R. Rivers, Instinct and the Unconscious, Cambridge University Press, 1920.

W. H. R. Rivers, The Influence of Alcohol and Other Drugs on Fatigue, Arnold, London, 1908.

W. McDougall and M. Smith, The Effects of Alcohol and Some Other Drugs During Normal and Fatigued Conditions, Privy Council, Medical Research Council, Special Report Series, No. 56, London, 1920.

W. McDougall, *Outline of Abnormal Psychology*, Chas. Scribner's Sons, 1926, p. 28.

W. McDougall, Outline of Abnormal Psychology, Chas. Scribner's Sons, 1926, p. 36.

W. McDougall, Outline of Abnormal Psychology, Chas. Scribner's Sons, 1926, p. 45.

W. McDougall, Outline of Abnormal Psychology, Chas. Scribner's Sons, 1926, pp. 28, 440.

W. McDougall, Outline of Abnormal Psychology, Chas. Scribner's Sons, 1926.

W. McDougall, Outline of Abnormal Psychology, Chas. Scribner's Sons, 1926, p. 441.

W. McDougall, Social Psychology, Luce & Co., (13th ed., 1918, p. 33.

W. McDougall, Social Psychology, Lute & Co., (13th ed.) 1918.

W. McDougall, supra, especially Ch. II.

●●●